The History of Life: A Very Short Introduction

VERY SHORT INTRODUCTIONS are for anyone wanting a stimulating and accessible way in to a new subject. They are written by experts, and have been published in more than 25 languages worldwide.

The series began in 1995, and now represents a wide variety of topics in history, philosophy, religion, science, and the humanities. Over the next few years it will grow to a library of around 200 volumes – a Very Short Introduction to everything from ancient Egypt and Indian philosophy to conceptual art and cosmology.

Very Short Introductions available now:

AFRICAN HISTORY
    John Parker and Richard Rathbone
AMERICAN POLITICAL PARTIES
    AND ELECTIONS L. Sandy Maisel
THE AMERICAN PRESIDENCY
    Charles O. Jones
ANARCHISM Colin Ward
ANCIENT EGYPT Ian Shaw
ANCIENT PHILOSOPHY Julia Annas
ANCIENT WARFARE
    Harry Sidebottom
ANGLICANISM Mark Chapman
THE ANGLO-SAXON AGE John Blair
ANIMAL RIGHTS David DeGrazia
ANTISEMITISM Steven Beller
ARCHAEOLOGY Paul Bahn
ARCHITECTURE Andrew Ballantyne
ARISTOTLE Jonathan Barnes
ART HISTORY Dana Arnold
ART THEORY Cynthia Freeland
THE HISTORY OF ASTRONOMY
    Michael Hoskin
ATHEISM Julian Baggini
AUGUSTINE Henry Chadwick
AUTISM Uta Frith
BARTHES Jonathan Culler
BESTSELLERS John Sutherland
THE BIBLE John Riches
THE BRAIN Michael O'Shea
BRITISH POLITICS Anthony Wright
BUDDHA Michael Carrithers
BUDDHISM Damien Keown
BUDDHIST ETHICS Damien Keown
CAPITALISM James Fulcher
CATHOLICISM Gerald O'Collins
THE CELTS Barry Cunliffe
CHAOS Leonard Smith

CHOICE THEORY Michael Allingham
CHRISTIAN ART Beth Williamson
CHRISTIANITY Linda Woodhead
CITIZENSHIP Richard Bellamy
CLASSICS Mary Beard and
    John Henderson
CLASSICAL MYTHOLOGY
    Helen Morales
CLAUSEWITZ Michael Howard
THE COLD WAR Robert McMahon
CONSCIOUSNESS Susan Blackmore
CONTEMPORARY ART
    Julian Stallabrass
CONTINENTAL PHILOSOPHY
    Simon Critchley
COSMOLOGY Peter Coles
THE CRUSADES Christopher Tyerman
CRYPTOGRAPHY
    Fred Piper and Sean Murphy
DADA AND SURREALISM
    David Hopkins
DARWIN Jonathan Howard
THE DEAD SEA SCROLLS
    Timothy Lim
DEMOCRACY Bernard Crick
DESCARTES Tom Sorell
DESIGN John Heskett
DINOSAURS David Norman
DOCUMENTARY FILM
    Patricia Aufderheide
DREAMING J. Allan Hobson
DRUGS Leslie Iversen
THE EARTH Martin Redfern
ECONOMICS Partha Dasgupta
EGYPTIAN MYTH Geraldine Pinch
EIGHTEENTH-CENTURY BRITAIN
    Paul Langford

THE ELEMENTS Philip Ball
EMOTION Dylan Evans
EMPIRE Stephen Howe
ENGELS Terrell Carver
ETHICS Simon Blackburn
THE EUROPEAN UNION John Pinder
    and Simon Usherwood
EVOLUTION Brian and
    Deborah Charlesworth
EXISTENTIALISM Thomas Flynn
FASCISM Kevin Passmore
FEMINISM Margaret Walters
THE FIRST WORLD WAR
    Michael Howard
FOSSILS Keith Thomson
FOUCAULT Gary Gutting
FREE WILL Thomas Pink
THE FRENCH REVOLUTION
    William Doyle
FREUD Anthony Storr
FUNDAMENTALISM Malise Ruthven
GALAXIES John Gribbin
GALILEO Stillman Drake
GAME THEORY Ken Binmore
GANDHI Bhikhu Parekh
GEOGRAPHY John A. Matthews and
    David T. Herbert
GEOPOLITICS Klaus Dodds
GERMAN LITERATURE Nicholas Boyle
GLOBAL CATASTROPHES Bill McGuire
GLOBALIZATION Manfred Steger
GLOBAL WARMING Mark Maslin
THE GREAT DEPRESSION AND
    THE NEW DEAL Eric Rauchway
HABERMAS James Gordon Finlayson
HEGEL Peter Singer
HEIDEGGER Michael Inwood
HIEROGLYPHS Penelope Wilson
HINDUISM Kim Knott
HISTORY John H. Arnold
THE HISTORY OF LIFE
    Michael Benton
THE HISTORY OF MEDICINE
    William Bynum
HIV/AIDS Alan Whiteside
HOBBES Richard Tuck
HUMAN EVOLUTION Bernard Wood
HUMAN RIGHTS Andrew Clapham
HUME A. J. Ayer
IDEOLOGY Michael Freeden
INDIAN PHILOSOPHY Sue Hamilton
INTELLIGENCE Ian J. Deary

INTERNATIONAL MIGRATION
    Khalid Koser
INTERNATIONAL RELATIONS
    Paul Wilkinson
ISLAM Malise Ruthven
JOURNALISM Ian Hargreaves
JUDAISM Norman Solomon
JUNG Anthony Stevens
KABBALAH Joseph Dan
KAFKA Ritchie Robertson
KANT Roger Scruton
KIERKEGAARD Patrick Gardiner
THE KORAN Michael Cook
LAW Raymond Wacks
LINGUISTICS Peter Matthews
LITERARY THEORY Jonathan Culler
LOCKE John Dunn
LOGIC Graham Priest
MACHIAVELLI Quentin Skinner
THE MARQUIS DE SADE John Phillips
MARX Peter Singer
MATHEMATICS Timothy Gowers
THE MEANING OF LIFE
    Terry Eagleton
MEDICAL ETHICS Tony Hope
MEDIEVAL BRITAIN
    John Gillingham and Ralph A. Griffiths
MEMORY Jonathan K. Foster
MODERN ART David Cottington
MODERN CHINA Rana Mitter
MODERN IRELAND Senia Pašeta
MOLECULES Philip Ball
MORMONISM
    Richard Lyman Bushman
MUSIC Nicholas Cook
MYTH Robert A. Segal
NATIONALISM Steven Grosby
NELSON MANDELA Elleke Boehmer
THE NEW TESTAMENT AS
    LITERATURE Kyle Keefer
NEWTON Robert Iliffe
NIETZSCHE Michael Tanner
NINETEENTH-CENTURY BRITAIN
    Christopher Harvie and
    H. C. G. Matthew
NORTHERN IRELAND
    Marc Mulholland
NUCLEAR WEAPONS
    Joseph M. Siracusa
THE OLD TESTAMENT
    Michael D. Coogan
PARTICLE PHYSICS Frank Close

PAUL E. P. Sanders
PHILOSOPHY Edward Craig
PHILOSOPHY OF LAW
   Raymond Wacks
PHILOSOPHY OF SCIENCE
   Samir Okasha
PHOTOGRAPHY Steve Edwards
PLATO Julia Annas
POLITICAL PHILOSOPHY
   David Miller
POLITICS Kenneth Minogue
POSTCOLONIALISM Robert Young
POSTMODERNISM Christopher Butler
POSTSTRUCTURALISM
   Catherine Belsey
PREHISTORY Chris Gosden
PRESOCRATIC PHILOSOPHY
   Catherine Osborne
PSYCHIATRY Tom Burns
PSYCHOLOGY
   Gillian Butler and Freda McManus
THE QUAKERS Pink Dandelion
QUANTUM THEORY
   John Polkinghorne
RACISM Ali Rattansi
RELATIVITY Russell Stannard
RELIGION IN AMERICA Timothy Beal
THE RENAISSANCE Jerry Brotton
RENAISSANCE ART
   Geraldine A. Johnson
ROMAN BRITAIN Peter Salway
THE ROMAN EMPIRE
   Christopher Kelly
ROUSSEAU Robert Wokler
RUSSELL A. C. Grayling
RUSSIAN LITERATURE Catriona Kelly
THE RUSSIAN REVOLUTION
   S. A. Smith

SCHIZOPHRENIA
   Chris Frith and Eve Johnstone
SCHOPENHAUER
   Christopher Janaway
SCIENCE AND RELIGION
   Thomas Dixon
SCOTLAND Rab Houston
SEXUALITY Véronique Mottier
SHAKESPEARE Germaine Greer
SIKHISM Eleanor Nesbitt
SOCIAL AND CULTURAL
   ANTHROPOLOGY
   John Monaghan and Peter Just
SOCIALISM Michael Newman
SOCIOLOGY Steve Bruce
SOCRATES C. C. W. Taylor
THE SPANISH CIVIL WAR
   Helen Graham
SPINOZA Roger Scruton
STATISTICS David Hand
STUART BRITAIN John Morrill
TERRORISM Charles Townshend
THEOLOGY David F. Ford
THE HISTORY OF TIME
   Leofranc Holford-Strevens
TRAGEDY Adrian Poole
THE TUDORS John Guy
TWENTIETH-CENTURY BRITAIN
   Kenneth O. Morgan
THE UNITED NATIONS
   Jussi M. Hanhimäki
THE VIETNAM WAR
   Mark Atwood Lawrence
THE VIKINGS Julian Richards
WITTGENSTEIN A. C. Grayling
WORLD MUSIC Philip Bohlman
THE WORLD TRADE
   ORGANIZATION Amrita Narlikar

Available Soon:

APOCRYPHAL GOSPELS Paul Foster
EXPRESSIONISM Katerina Reed-Tsocha
FREE SPEECH Nigel Warburton
MODERN JAPAN
   Christopher Goto-Jones

NOTHING Frank Close
PHILOSOPHY OF RELIGION
   Jack Copeland and Diane Proudfoot
SUPERCONDUCTIVITY
   Stephen Blundell

For more information visit our websites
www.oup.com/uk/vsi
www.oup.com/us

Michael J. Benton

THE HISTORY
OF LIFE

A Very Short Introduction

OXFORD
UNIVERSITY PRESS

OXFORD
UNIVERSITY PRESS

Great Clarendon Street, Oxford OX2 6DP

Oxford University Press is a department of the University of Oxford.
It furthers the University's objective of excellence in research, scholarship,
and education by publishing worldwide in

Oxford New York

Auckland Cape Town Dar es Salaam Hong Kong Karachi
Kuala Lumpur Madrid Melbourne Mexico City Nairobi
New Delhi Shanghai Taipei Toronto

With offices in

Argentina Austria Brazil Chile Czech Republic France Greece
Guatemala Hungary Italy Japan Poland Portugal Singapore
South Korea Switzerland Thailand Turkey Ukraine Vietnam

Oxford is a registered trade mark of Oxford University Press
in the UK and in certain other countries

Published in the United States
by Oxford University Press Inc., New York

British Library Cataloguing in Publication Data
Data available

Library of Congress Cataloging in Publication Data
Data available

ISBN 978-0-19-922632-0

1 3 5 7 9 10 8 6 4 2

Typeset by SPI Publisher Services, Pondicherry, India
Printed in Great Britain by
Ashford Colour Press Ltd, Gosport, Hampshire

Contents

List of illustrations ix

Introduction 1

1 The origin of life 15

2 The origin of sex 33

3 The origin of skeletons 51

4 The origin of life on land 69

5 Forests and flight 87

6 The biggest mass extinction 101

7 The origin of modern ecosystems 122

8 The origin of humans 146

Index 167

List of illustrations

1 A selection of fossils from
   a mid-Victorian textbook
   (1860) **3**
   Mansell/Time & Life Pictures/
   Getty Images

2 An exceptionally well
   preserved fossil from Liaoning
   Province, China **6**
   Spencer Platt/Getty Images

3 Geological timescale **18–19**

4 The formation of an RNA
   protocell **28**
   Reprinted by permission from
   Macmillan Publishers Ltd (*Nature*
   2001)

5a Stromatolite fossils in the
   Stark Formation, Mackenzie,
   Canada **30**
   P. F. Hoffman (GSC)

5b Filamentous microfossils in a
   3,235-million-year-old
   massive sulfide from
   Australia **31**
   Courtesy of Birger Rasmussen

6 The universal tree of life **36**
   Professor Norman Pace

7 The endosymbiotic theory for
   the origin of eukaryotes **38**
   Inspired by www.thebrain.mcgill.ca

8 A close up of *Bangiomorpha*
   filaments **46**
   Dr Nick Butterfield

9 Life as it may have looked in
   Ediacaran times **49**
   Smithsonian Institution

10 Fossils from the Early
   Cambrian **55**
   M. Alan Kazlev/Dorling Kindersley

11 The Burgess Shale scene,
   Middle Cambrian **58**
   Christian Jegou Publiphoto
   Diffusion/Science Photo Library

12 *Cooksonia* **74**

13 The Rhynie ecosystem **76**
   Simon Powell, Bristol University

14 *Ichthyostega* and *Acanthostega* reconstructions **80**
Mike Coates

15 A Carboniferous riverbank **88**
Walter Myers

16 Life before and after the end-Permian mass extinction **104**
John Sibbick

17 Life on land in the Late Permian in what is now Russia **110**
John Sibbick

18 The pattern of marine extinction through the end-Permian crisis **114**
From fig. 1, Y. G. Jin et al., *Science* 289: 432–36 (21 July 2000). Reprinted with permission from AAAS

19 Reptiles from the Triassic **127**
From Mike Benton, *Vertebrate Palaeontology* (3rd edn., Blackwell, Oxford, 2005)

20 Dinosaurs of the Late Jurassic of North America **139**
Ernest Unterman/Dinosaur National Monument Museum, Utah/Bettmann/Corbis

The publisher and the author apologize for any errors or omissions in the above list. If contacted they will be pleased to rectify these at the earliest opportunity.

Introduction

> The Age of Reptiles ended because it had gone on long enough and
> it was all a mistake in the first place.
>
> Will Cuppy, *How to become extinct* (1941)

It is hard to make sense of the history of life on Earth. A mass of
strange and extraordinary animals and plants perhaps flits before
our eyes when we think of prehistory: Neanderthal man,
mammoths, dinosaurs, ammonites, trilobites ... and of course a
time when there was no life at all, or at least merely microscopic
beasts of extreme simplicity floating in the primeval ocean.

These impressions come from many sources. Children today are
weaned on dinosaur books, and the images of living, breathing
dinosaurs are everywhere, in movies and television
documentaries. Then, too, as children, many people have gone to
coastal cliffs or quarries and collected their own fossil ammonites
or trilobites. These common fossils, as well as many much more
spectacular and beautiful examples, such as petrifactions of
exquisite fishes showing all their scales, still shiny after millions of
years, may be seen in fossil shops, or in lavish photographs in
coffee table books and on the web.

Most people are aware that dinosaurs, despite their ubiquity in
modern culture, lived a long time before the first humans, and

there were untold spans of time before the dinosaurs existed that were populated by ever-more unusual and strange animals and plants. How are we to make sense of all of this?

Fossils

The keys to understanding the history of life are fossils (Fig. 1). *Fossils* are the remains of plants, animals, or microbes that once existed. Fossils may be *petrifactions*, which means literally 'turned into rock', and these are some of the commonest examples. Petrified fossils may be of two kinds, first, those that are literally turned to rock, and where none of the original organism remains. The leaf or tree trunk, or shell, or worm, has completely disappeared, and the cavity left behind has been replaced by grains of sand or mud, or more often by minerals in solution that have flowed through the spaces in the surrounding rock and have then infiltrated the space and crystallized.

The second, and commoner, kind of petrifaction still retains some of the original material of the animal, perhaps the calcium carbonate that made up the shell, or some cuticle or carbonized relic of the plant. Rock grains or minerals then merely fill the cavities. So, many people might be surprised to realize that common fossils, such as a 400-million-year-old trilobite or a 200-million-year-old ammonite, are actually largely made from the original calcium carbonate of their external skeleton or shell, as in life. Similarly, by far the majority of dinosaur bones are still made of the original calcium phosphate (apatite), the main mineralized constituent of bone then and today. If you look closely at the outer surface of these fossils, perhaps with a magnifying glass, you can see extremely fine features, such as pimples and growth lines on the trilobite carapace, original multicoloured mother-of-pearl on the ammonite shell, and muscle scars or tooth marks on the surface of the dinosaur bone. If the fossil shells or bones are cut across and examined under the microscope, all the original growth layers and internal structures are still there. So, a

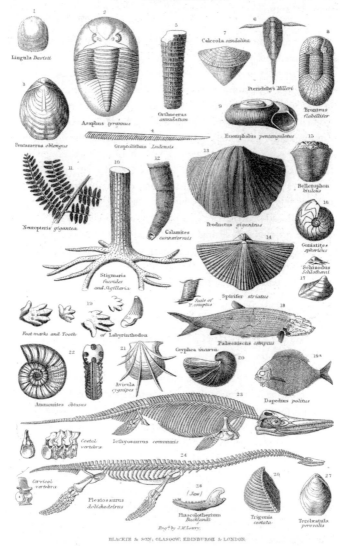

1. A selection of fossils from a mid-Victorian textbook, showing trilobites (top), Coal Measure plants and brachiopods (centre) and a selection of ammonites, fossil fishes, an ichthyosaur, and a plesiosaur (lower half)

section cut through dinosaur bone looks just as fresh today as a section through a modern bone.

Every plant or animal that has ever lived has not turned into a fossil. Indeed, if this were the case, the surface of the Earth would be covered in avalanches of fossils everywhere, great mounds of dinosaur bones, trilobites, giant coal forest trees, ammonites, and the like, probably extending to the moon. No one knows what proportion of life has ended up fossilized, but it is clearly a tiny fraction, much less than 1 per cent. Plants or animals must at least have *hard parts* such as a skeleton, a shell, or a toughened, woody trunk to be readily preservable. Even so, the majority of animal carcasses and dead plants enter the food chain almost immediately, being scavenged by animals or decomposed by bacteria. Dead organisms can only turn into fossils if *sedimentation* is happening, that is, sand or mud are being dumped on top of the remains, perhaps on the floor of a deep lake, under a sand bar in a river, or deep in the ocean, below the zone that is constantly churned up by currents and tides.

Worms and feathered dinosaurs: exceptional preservation

Other fossils may be preserved in slightly unusual conditions that may, on occasion, provide unique and unexpected insights into ancient life, so-called *exceptional preservation*. Exceptionally preserved fossils may show soft structures, such as flesh, eyes, stomach contents, feathers, hair, and the like. Sites of exceptional fossil preservation are sometimes called 'windows' on the life of the past. They allow *palaeontologists*, the scientists who study fossils, to see a snapshot of everything that existed at particular times and in particular places. These at least allow palaeontologists to see the soft-bodied worms, jellyfish, and other creatures that are rarely preserved in normal circumstances.

The Burgess Shale in Canada is one of the most famous of these sites of exceptional preservation. These rocks are 505 million years old, so they document some of the oldest animals. Without the Burgess Shale, and similar sites of about the same age in Greenland and China, palaeontologists would know only about shelled and skeletonized organisms such as brachiopods ('lamp shells'), trilobites, and sponges. The Burgess Shale has increased our knowledge of life in the Cambrian many-fold: it has revealed whole clans of worm-like creatures, some related to modern swimming and burrowing worms, others seemingly unique and hard to link to modern animals. The Burgess Shale also shows the feathery legs and gills of the trilobites, their mouths, guts, and sense organs, and it reveals strange tadpole-like swimming animals that have primitive backbones and so are close to our own ancestry.

Equally famous are the sites of exceptional preservation in Liaoning Province in north-east China. These date back to 125 million years ago, and they have produced spectacular fossils of birds (and dinosaurs) with feathers and internal organs, mammals with hair, fishes with gills and guts, and any number of worms, jellyfish, and other soft-bodied denizens of those ancient Chinese lakes (Fig. 2).

There are dozens of other such sites of exceptional preservation scattered pretty randomly through time and space. But why do they exist and how are the soft structures preserved? Most of these sites come from times and places where oxygen was limited. Deep lakes and deep oceans sometimes lose the normal oxygen content of the waters, if, for example, there is a dramatic growth of algae and other floating plants at the surface, a so-called algal bloom. These occur in warm conditions, and the lakes and oceans may become temporarily stagnant. The stagnation of the waters may itself kill swimming creatures, and beasts that crawl around on the bottom muds. The lack of oxygen can also mean that the normal

2. An exceptionally well preserved small dinosaur specimen, *Microraptor*, **from the Early Cretaceous of Liaoning Province, China**

scavenging creatures cannot survive, and the carcasses do not have all their flesh stripped.

Experiments show that, in oxygen-poor, or *anoxic*, conditions, soft tissues, even muscles, guts, and eyeballs, can be invaded by minerals that come from the body fluids of the animals, or from the surrounding sediments. These are typically flash-mineralizing processes, where the fibres of a muscle, or the complex tissues of a gill or a stomach, are invaded and replaced within hours or days at most. Once mineralized, the replicas of soft tissues can then survive to the present day.

Living blimps? Quality of the record

Like most palaeontologists, I sometimes sit bolt upright in bed at night and worry whether the fossil record is informative or not. Charles Darwin wrote about the 'imperfection of the geological record', and he was well aware that most organisms are never fossilized, and so palaeontologists miss so much of ancient life.

The question though is: how much is missing? Is it 50 per cent or 90 per cent or 99.99999 per cent? This can never be determined, of course. A more sensible question might be: how adequate is the fossil record?

Palaeontologists have speculated that there might be whole sectors of extinct life that we know nothing about. What if there were a diverse class of floating animals that were constructed of extremely lightweight materials, and provided with great air bladders that filled with gases lighter than air? These creatures might have been many metres long, perhaps as large as dirigible aircraft, sometimes called blimps during the Second World War. These blimp beasts could well have dominated the Earth, if they were so large, and yet they might have entirely escaped fossilization. Their bodily tissues might have been so lightweight that they rotted away when they died. Their gas bladders would clearly burst and disappear during decay. Living in the air, in any case, means their carcasses might have generally fallen onto the surface of the Earth, and so they might not often have been covered with sediment in any case.

Palaeontologists have no way of detecting such hypothetical extinct beasts. Other soft-bodied creatures can be assumed to have existed, though. For example, there are many *phyla*, or major groups, of worm-like creatures today, nematodes, platyhelminths, gastrotrichs, sipunculids, and others, that have no known fossil representatives. And yet, because they exist today, and because we can establish their evolutionary relationships to other organisms with shells or skeletons, we know the length of their missing fossil record. If a soft-bodied worm group is the closest relative of another wormy creature with a shell, both groups must have existed for the same length of time; their common ancestor must have lived at a particular time, and the fossil record of the shelled group establishes a minimum age for both groups. The known missing record of the soft-bodied group is called a *ghost range*, a part of the missing fossil record we can predict with some certainty.

What do the sites of exceptional preservation tell us? If they preserve more or less everything that lived at the time, soft- and hard-bodied, they can be used as a yardstick against which to test the 'normal' fossil record. It seems that the ancient exceptional sites, such as the Burgess Shales, tell us more about unknown groups than the more recent ones, such as the Liaoning beds in China. In fact the soft-bodied organisms from Liaoning, worms, jellyfish, insects, and the like, are all entirely predictable from other known fossils and from ghost ranges.

Palaeontologists have been pretty assiduous in retrieving fossils. As time goes on, it now seems to take much more effort than it took a century ago to find something new. Indeed, not much has changed in our knowledge of the fossil record since the time of Darwin. In the 1850s, palaeontologists knew about trilobites and ammonites, fossil fishes, dinosaurs, and fossil mammals. They did not know anything about the first life from the Precambrian, nor did they know much about human evolution. But the fact that neither trilobites nor humans have been found in the age of the dinosaurs, nor have any other fossils been found in seriously unexpected places, suggests that the record is known more or less well. Our work now is merely to flesh out the details.

But that still says nothing about the giant blimps...

Molecules and the history of life

It might seem unexpected to introduce molecular biology at this point. But, just as historians have parallel sets of evidence from artefacts and from written records, so too do students of the history of life. Until the 1960s, there were only fossils; after that there were also molecules – even though most palaeontologists at the time probably did not appreciate it.

In an extraordinary paper published in 1962 by Emil Zuckerkandl and Linus Pauling, in a rather obscure conference volume, the *molecular clock* was born. Molecular biology had arisen ten years earlier when, in 1953, James Watson and Francis Crick announced the structure of deoxyribose nucleic acid, DNA, the chemical that makes up genes and is the basis of the genetic code. By 1963, several proteins, such as haemoglobin, the protein that carries oxygen in the blood and makes it red, had been sequenced, that is, the detailed structure had been determined, and the new breed of molecular biologists had noted something extraordinary. The proteins of different species of animal were not identical, and their structures differed more between distantly related species. In other words, the haemoglobin molecules of humans and chimpanzees were identical, but the haemoglobin of a shark was very different.

Zuckerkandl and Pauling took the brave leap of suggesting, on rather limited evidence then, that the amount of difference was proportional to time. The negligible difference between the haemoglobins of humans and chimpanzees showed these two species had diverged only a short time ago, geologically speaking, whereas the 79 per cent difference between human and shark haemoglobin pointed to a divergence 400 million years ago, or more.

In the 1960s, protein sequencing was a laborious process, and the new data came slowly, but by 1967 the haemoglobin of the great apes was known sufficiently that the first attempt was made to produce an evolutionary tree. The science of molecular phylogenetics was born. Vincent Sarich and Allan Wilson, in a three-page paper in the American journal *Science*, plotted the relationships of humans and apes, and showed that our nearest relative was the chimpanzee, then the gorilla, and then the orang-utan. This was not so unexpected, and it agreed with the pattern of relationships established from studies of anatomy. The shocking part of the paper was that the molecular clock said humans and chimps had diverged only 5 million years ago.

Palaeontologists were variously bemused and horrified. Most dismissed the new technique: after all, if it produced such ludicrous results, it was clearly not working. Everyone knew that humans and chimpanzees had split some 15–20 million years ago, based on studies of *Proconsul* and other early human-like fossils from the Miocene of Africa. Others took the method seriously, but were equally unhappy about the result.

As the protein data sets grew, more mammals were added to the tree, and the branching dates seemed quite reasonable for most other groups. This increased the nervousness of the palaeontologists, who then faced a conundrum: do we accept the new molecular date, or insist on the established fossil evidence? Slowly, they came to realize the molecular date was probably right. Closer study of the fossils showed that they had been over-interpreted. The supposedly 'human' characters of *Proconsul* and its kin were not really human at all. This fossil was related to the common ancestors of humans and the African apes, and so said nothing about the true timing of divergence. Since the 1970s, new finds in Africa have shown that the divergence date between humans and chimps must be at least 6–7 million years ago.

Now, molecular biologists interested in the *tree of life*, the great pattern of relationships linking all species, use DNA sequences. Protein sequencing is slow, and the evidence limited. DNA, the genetic code, offers much more information, and new techniques developed in the 1980s have made sequencing almost automatic. Computers can also crunch enormous masses of data these days, so sequencers are happy to run lengthy segments of the genetic code, consisting of many genes, and for dozens, or even hundreds of species, to produce patterns of relationships for specific groups or for large sectors of life. It is possible to assess the genome of, say, twenty species of lizards, and draw up a tree that documents evolution over a span of perhaps 10 million years. Equally, the analyst can select, say, twenty species across all of life – a human, a shark, a mollusc, a tree, a fern, a bacterium – and find a tree of relationships that extends deeply back in time.

But where do the fossils fit into all this?

Cladistics

I remember when I attended my first scientific meeting, as an undergraduate, a session of the Society for Vertebrate Palaeontology and Comparative Anatomy at University College, London, in 1976, I wondered if I would ever go back. As I looked on nervously, the big beasts of the subject were bickering and squabbling appallingly over something called 'cladistics'. I'd heard nothing about this – it wasn't taught then as part of my degree. One person would assert with fervour that everyone should adopt this new technique. Another would say it was all nonsense – even a Marxist plot to overthrow the scientific method. I stumbled back to the train, wondering whether my decision to become a professional palaeontologist was mistaken. Were they all mad?

On reading around, I discovered that cladistics had been promulgated by a German entomologist, Willi Hennig. He had written about the technique in the 1950s, but it had only really attracted attention when the book was translated into English and reissued in 1966. But, from 1966 to 1980, only a rather small group of true believers espoused the method, and it had not in any way become mainstream. Hennig argued passionately that *systematists*, the biologists and palaeontologists who were interested in species and the tree of life, should be more objective in their methods.

Until Hennig's time, systematists had attempted to draw up trees of relationships based on a judicious sifting of the character evidence. A biological *character* is any observable feature of an organism – 'possession of feathers', 'possession of four fingers', 'iridescent blue feathers on top of the head', 'multiple flower heads on each stem' – and systematists had long understood that if two organisms share a character they might well be related. The problem was always *convergence*, the well-known observation that unrelated organisms might evolve similar features independently. Insects, birds, and bats have wings, but no one ever suggested that this was sufficient evidence to group these organisms together as close relatives: in detail, their wings are anatomically quite different in structure, and so they evolved them independently, but for the same purpose. But how were systematists to distinguish convergence from truly shared, evolutionarily identical, characters?

This was Hennig's point: objective techniques were required to distinguish truly shared characters from convergences, but also to distinguish inherited 'primitive' characters from those that truly marked a particular branching point. So, while it is true that humans and chimpanzees share the character 'hand with five fingers', and this is not a convergence, the character is not helpful at the level of the branching point between the two species. In fact, all land-living vertebrates basically have a five-fingered hand – lizards, crocodiles, dinosaurs, rats, bats, whales, and so on. Hennig had identified the critical point, that anatomical characters had to be evolutionarily unique (not convergent) and they had to be assessed at the correct level in the tree before they could be considered useful. He termed such characters *synapomorphies*, sometimes rendered in English as 'shared derived characters'. (Hennig's writing, in any language, is heavy going, and he liked inventing long words – neither of which helped gain him converts.)

Hennig's concept of a synapomorphy is more or less the same as the classic notion of a *homology*, that is, any structure that shares a common fundamental pattern because of common ancestry – such as the human arm, the wing of a bat, and the paddle of a whale. These limbs may have different functions today, but they all share the same bones and muscles inside, and we now know they evolved from the ancestral front limb of the first mammal.

Since the 1970s, systematists have increasingly switched to using cladistics in their work. After all, there was no alternative – the older techniques were really inspired guesswork. Acceptance came largely for a reason Hennig could not have predicted, namely the growth in power and ease of use of computers. The secret to cladistics is the *character matrix*, a listing of all the species of interest, and codings of their characters (1 for presence, 0 for absence). Multiple cross-checking over the matrix, and repeated runs of the analysis, provided statistical methods of assessing which tree or trees explained the data best, and the probability that synapomorphies were correctly identified or not. In practice, there have been many problems, but cladistic methods are ubiquitous, and repeat analyses by different analysts allow published trees to be tested and confirmed or rejected.

The great leap forward

Palaeontologists are aware that their field has transformed itself immeasurably since the 1960s, but public attention has focused elsewhere – the space race, genetic engineering, computer technology, nanoscience, global change. But, cladistics and molecular phylogeny have introduced new rigour into the field of drawing up evolutionary trees. Whereas in the 1950s and 1960s a palaeontologist did his or her best to make a tree by 'joining the dots' – linking similar-looking beasts through time – today there are many independently derived trees of the evolution of different groups, some based on different genes, others on different combinations of fossil and recent data on anatomy. But do they agree?

The astonishing discovery is that molecular and palaeontological trees agree with each other more often than not. The two approaches are pretty well independent, so it is possible then to compare, say, a tree based on molecular sequences of modern rodents with a tree constructed by measuring the teeth and other anatomical features of living and extinct species. Inevitably, everyone hears about the cases where the results disagree. In the early days of molecular sequencing, some bizarre results emerged, but the methods were young, and mistakes were easy to make. Such bizarre results are rare now. In some cases, palaeontologists have humbly accepted that they have been entirely unable to resolve certain parts of the evolutionary tree, and the molecules give an unequivocal answer straight away. In other cases, there is no resolution yet, and more work is required. Some parts of the great tree of life may remain forever mysterious, perhaps because rates of evolution were so fast that characters did not accumulate, or the branching points are so ancient that subsequent evolution has obliterated the clues to relationship.

The third methodological or technological advance has been in dating the rocks. Since the 1960s, the accuracy of dating has improved greatly, and sequences of rocks and sequences of events can be compared more accurately than before. But we can look at that later. Let's begin the story.

Chapter 1
The origin of life

As a general rule, then, all testaceans grow by spontaneous generation in mud, differing from one another according to the differences
of the material; oysters growing in slime, and cockles and the other
testaceans above mentioned on sandy bottoms; and in the hollows of
the rocks the ascidian and the barnacle, and common sorts, such as
the limpet and the nerites.

Aristotle, *History of Animals*

From the earliest days people have wondered about the origins of
life. The ancient Greeks and Romans considered the topic, and
had many ideas. Most, like Aristotle (384–322 BC), focused on the
idea of *spontaneous generation*, a process that they believed
happened today, and that had presumably happened when life
first arose. As Aristotle wrote above, he believed that marine
shellfish all arose spontaneously from the mud, sand, and slime on
the seabed and among the coastal rocks. He made similar
assumptions about other forms of life: moths arose from woollen
garments, garden insects arose from the spring dew or from
decaying wood, and many fishes arose from froth on the surface of
the ocean. Such views held sway until the nineteenth century.

Louis Pasteur (1822–95) famously showed conclusively that life
could not arise spontaneously. He repeated experiments that had
been performed before, but took great pains to exclude all

possibility of contamination. Earlier workers had gone through the process of boiling a broth of water and hay in sealed flasks so that anything living in the water or the air within the flasks would be killed. But, despite these precautions, they still found microscopic organisms living in the water, and Pasteur argued that the germs entered the vessels when they were being cooled in a mercury trough. So he repeated the experiments, sterilizing the glassware and the water in the flasks, but ensuring also that laboratory air could not enter the cooling mixtures. With the air excluded, nothing living was detected in the boiled water even many months later.

The age of the Earth

The death of spontaneous generation was not the only problem for scientists interested in studying the origin of life about 1900. They also had no truly ancient fossils to work with, and no real idea of the age of the Earth, nor of the major events that might have preceded the origin of life. There was a widely held view that the Earth was something like a huge ball of iron – iron is one of the commonest elements – that had once been molten, and had been cooling down. Indeed, the eminent late Victorian physicist William Thomson, later Lord Kelvin (1824–1907), used this assumption, and his knowledge of thermodynamics, to speculate that the Earth formed only 20–40 million years ago.

Kelvin's view that the Earth was relatively young influenced many people at the turn of the twentieth century. No matter that the biologists and geologists were quite unhappy with this estimate; the leading physicist of the day had pronounced, and he had based his evidence on clear calculations. Charles Darwin had long assumed, for example, that the Earth must be hundreds or thousands of millions of years old, although he never speculated more closely than that. Nonetheless, he could see how the rocks of the south coast of England had accumulated rather slowly, made up from many millions of thin layers, each perhaps representing a year or a century. Other geologists held similar views, whether based on their calculations of the time taken for sedimentary rocks, such as limestones and mudstones, to accumulate, or the time it might have taken for the oceans to separate from the initial molten rock, and then to become salty.

Ironically, Kelvin lived through the crucial discoveries that were to show that his physical view of the Earth was too simplistic, but he was reluctant to shift. The discovery of radioactivity by Henri Becquerel (1852–1908) in 1896, the property of certain elements, such as uranium, radium, and polonium, to emit rays and to change their atomic number, changed everything. *Radioactive* elements may *decay* into another element, with the emission of rays. In radioactive decay, the *parent* element, such as uranium, would decay into another element, called the *daughter*, such as thorium, over a certain amount of time.

The discovery of radioactivity caused excitement throughout the world of physics, and only four years later, Ernest Rutherford (1871–1937) and Frederick Soddy (1877–1956) showed that radioactive decay is *exponential* – that is, the quantity of radioactive material halves over fixed amounts of time. In other words, 1,000 atoms of uranium reduce to 500 in a certain span of time, those 500 to 250 in the same amount of time, then to 125, and so on. Three years later, and in the hearing of an ageing and somewhat crotchety Lord Kelvin, Ernest Rutherford suggested that radioactive decay might provide a geological clock. He argued that, if scientists measured the time it takes for half the quantity of the parent radioactive element to decay to the daughter element, a span since called the *half life*, measurements of the proportions of parent to daughter element in a suitable rock sample could then give an estimate of the age of the rock.

Rutherford's suggestion was put into practice remarkably rapidly. In a bravura performance, the young British geologist Arthur Holmes (1890–1965), aged only 21 at the time, published the first

Aeon	Era	Period	
Phanerozoic	Cenozoic	Quaternary	
		Neogene (Tertiary)	
		Palaeogene (Tertiary)	
	Mesozoic	Cretaceous	
		Jurassic	
		Triassic	
	Palaeozoic	Permian	
		Carboniferous	Pennsylvanian
			Mississippian
		Devonian	
		Silurian	
		Ordovician	
		Cambrian	
Precambrian	Proterozoic		
	Archaean		
	Hadean		

3. **Geological timescale**

Epoch		Origins
Holocene		
	0.01	
Pleistocene		*Homo*
	1.8	
Pliocene		Apes and humans
	5.3	
Miocene		
	23.0	
Oligocene		
	33.9	Modern orders of animals and plants
Eocene		
	55.8	
Palaeocene		
	65.0	
UPPER		Flowering plants
LOWER		
	145	
UPPER		Reptiles flourish
MIDDLE		Birds
LOWER		
	200	
UPPER		Cone-bearing plants
MIDDLE		Mammals, dinosaurs
LOWER		
	252	
UPPER		Mammal-like reptiles
LOWER		
	299	
UPPER		Seed plant forests
	318	Reptiles
LOWER		
	359	
UPPER		Amphibians, insects, plants
MIDDLE		
LOWER		
	416	
UPPER		Fishes
LOWER		
	444	
UPPER		Colonization of land
MIDDLE		
LOWER		
	488	
UPPER		Most modern phyla
MIDDLE		
LOWER		
	542	
		Soft-bodied animals
		Algae
	2500	
		Bacteria
	4000	
	4560	

age estimates for rocks in 1911: his estimated dates ranged from 340 million years (a Carboniferous rock), to 1,640 million years (a Precambrian rock). These are not far off the modern age estimates (Fig. 3). Note that the first nine-tenths of the history of the Earth is called the Precambrian, because it precedes the Cambrian period: this is rather an apologetic, or negative term, for such a vast span of the Earth's history, but the term is established now and cannot be readily changed.

After the first very crude estimates had been made, Holmes, and many others, worked hard to improve their understanding of age measurements, and the chemistry and physics were much revised, so that by 1927 Holmes was able to produce a reasonable summary of key dates for the history of the Earth. Holmes suggested that the age of the Earth was between 1,600 and 3,000 million years. In the same year, Rutherford suggested 3,400 million years, and by the 1950s, the age of the Earth was estimated at 4,500–600 million years, the currently accepted figure. It was, and still is, hard to date the exact origin of the Earth because rocks were presumably molten then, and so there are no solidified crystals that may be dated.

Making the Earth habitable

There is some debate about when the Earth became habitable: did it take 200 or 600 million years? Most geologists have favoured the latter view: after all the initially molten surface had to cool to below 100 °C, or any organic compounds would have been burnt off. Life is based on carbon, hydrogen, and oxygen, and these all remain in a gaseous state at high temperatures. Of course water boils at 100 °C, and life is essentially water (H_2O) with carbon.

The Sun and its accompanying planets formed some 4.6 billion years ago from gas into which earlier generations of stars had spewed not only hydrogen and helium but small amounts of carbon, oxygen, and other elements forged in their cores. At first, the Earth was a molten mass, but it cooled, separating into an outer cool crust and an inner molten mantle and core. The heavier iron sank to the core, while lighter elements such as silicon rose to the surface. It took some 50 million years for the separation to occur, and the Moon may have spun off at this time, the result, it is thought, of a collision with an enormous planetoid. Massive volcanic eruptions rent the semi-molten silicon-rich rocks at the Earth's surface, and produced great volumes of gases: carbon dioxide, nitrogen, water vapour, and hydrogen sulphide. Temperatures on the Earth's surface were too high, and the crust was too unstable, for any form of carbon-based life to exist. At this time, the record of craters on the Moon suggests that there were a few huge impacts on Earth, impacts from large comets or asteroids that would have provided enough energy to turn the ocean into steam. Thus, if life had got started before 4 billion years ago, it would probably have been wiped out, only to start afresh.

As the Earth's surface cooled, the *lithosphere*, the rocky crust and outer mantle, began to differentiate as a cooler upper layer above the underlying *asthenosphere*. As the rocky lithosphere formed, and the upper crust divided into plates that were moved by mantle convection, slow-moving gyres of heat rising from the depths of the mantle moved laterally as they came close to the base of the cooler solid crust, and began the stately journey of the Earth's tectonic plates.

Geologists keep searching for the oldest rocks on Earth, and they are at all times pushing the limits of what might be possible (molten rocks cannot be dated, and error bars on dates become quite large when such ancient dates are attempted).

The oldest rock unit on Earth is said to be the Acasta Gneiss from the Northwest Territories, Canada, dated at up to 4.0 billion years old. This is a metamorphic rock, and the date is assumed to reflect the age of the older granite from which the gneiss was formed. Even older are zircons, isolated mineral grains, from the Jack Hills in Australia, which have yielded a date of 4.4 billion years. Could these minerals really have been solid, and even accumulating under water, at that point? Their discoverers claim this is the case, while others are sceptical that the Earth could have been cool enough for water to exist so soon after its formation.

The oldest sedimentary rocks have been reported from the Isua Group in Greenland, dated at 3.8–3.7 billion years ago. There is no doubt that water existed on the Earth by this point, and that some of the Isua Group rocks really are formed from accumulated sand, laid down under water, and deriving from older rock sources. It has even been claimed that these oldest sedimentary rocks also contain traces of life, but this claim is still much debated.

Traces of early life

In 1996, Stephen Mojzsis, then a graduate student at the Scripps Institution of Oceanography at La Jolla, California, made a startling announcement in the journal *Nature*. He claimed to have identified a clear chemical signature for life in carbon compounds from Isua Group rocks. He had analysed minute grains of graphite, a form of carbon, in the rocks, and found an unusually high proportion of carbon-12. The carbon atom has two stable isotopes, carbon-12 and carbon-13. The ratio of these two forms of carbon can indicate the presence or absence of organic residues of previously living organisms: enrichment in carbon-12 relative to carbon-13 is characteristic of photosynthesizing organisms, and the organisms that eat them. Mojzsis was confident he had identified life: 'Our evidence establishes beyond reasonable doubt that life emerged on Earth at least 3.85 billion years ago, and this is not the end of the story. We may well find that life existed even earlier.'

If the interpretation is correct, then the grains of graphite in the Isua rocks prove that photosynthesis was happening 3.85 billion years ago. *Photosynthesis* is the process by which green plants convert energy from sunlight into food. Carbon dioxide and water combine, and produce oxygen, usually given off as a gas, and sugars, which form the building blocks of the plant. Now, in the early part of the history of the Earth, these photosynthesizing organisms were not trees or flowers, but presumably simple microbes known as cyanobacteria.

Other researchers have argued strongly against this interpretation. They noted, for example, that the Isua graphite was not in the sedimentary rocks of the area, but in the metamorphic rocks. Indeed, the Isua sedimentary rocks contained relatively low proportions of graphite. The alternative argument was then that the Isua graphites were of secondary, inorganic origin and might have formed by heating of iron carbonate. One of the critics, Roger Buick of the University of Washington, Seattle, said that 'These rocks have been buried and cooked at least three times. They've been severely squashed and strained and tied in knots at least three times too.'

The Isua graphites are still held as evidence for early life, and the debates continue to rage. But how does this chime with current theoretical views about the origin of life?

The biochemical theory for the origin of life

There are many models for the origin of life, all based on an understanding of how the simplest living organisms today operate. The first 'modern' model for the origin of life was presented in the 1920s independently by two remarkable scientists, the Russian biochemist A. I. Oparin (1894–1980) and the British evolutionary biologist J. B. S. Haldane (1892–1964). Oparin and Haldane share the distinction of being independent co-founders of the so-called biochemical theory for the origin of life, as well as being known normally only by their initials.

According to the Oparin–Haldane model, life could have arisen through a series of organic chemical reactions that produced ever more complex biochemical structures. They proposed that common gases in the early Earth atmosphere combined to form simple organic chemicals, and that these in turn combined to form more complex molecules. Then, the complex molecules became separated from the surrounding medium, and acquired some of the characters of living organisms. They became able to absorb nutrients, to grow, to divide (reproduce), and so on. The Oparin–Haldane model was not tested until the 1950s.

In 1953, Stanley Miller (1920–2007), then a student of Harold Urey (1893–1981) at the University of Chicago, made a model of the Precambrian atmosphere and ocean in a laboratory glass vessel. He exposed a mixture of water, nitrogen, carbon monoxide, and nitrogen to electrical sparks, to mimic lightning, and found a brownish sludge in the bottle after a few days. This contained sugars, amino acids, and nucleotides. So Miller had apparently recreated the first two steps in the Oparin–Haldane model, mixing the basic elements to produce simple organic compounds, and then combining these to produce the building blocks of proteins and nucleic acids.

It should be noted that critics have said that the mixture of gases that Miller used (with high percentage concentrations of hydrogen and methane) was rather different from the likely atmosphere of the early Earth. Atmospheric hydrogen is ultimately replenished from the mixture of gases released from the solid Earth; but the geochemistry of the subsurface means that the mixture generally should contain the oxidized form of hydrogen, namely water vapour, H_2O, rather than the large proportion of free hydrogen gas in Miller's model atmosphere.

Further experiments in the 1950s and 1960s led to the production of polypeptides, polysaccharides, and other larger organic molecules, the next step in the hypothetical sequence. Sidney Fox at Florida State University even succeeded in creating cell-like structures, in which a soup of organic molecules became enclosed in a membrane. His 'protocells' seemed to feed and divide, but they did not survive for long, so they were not living, despite the hype made by the press at the time.

In a recent twist to the classic Oparin–Haldane biochemical model, Euan Nisbet (University of London) and Norman Sleep (Stanford University) proposed the hydrothermal model for the origin of life in 2001. In this model, the ancestor of all living things was a hyperthermophile, a simple organism that lived in unusually hot conditions. The transition from isolated amino acids to DNA may then have happened in a hot-water system associated with active volcanoes, rather than in some primeval soup at the ocean surface. There are two main kinds of hot-water systems on Earth today, 'black smokers' found in the deep oceans above mid-ocean ridges where magma meets sea water, and hot pools and fumaroles fed by rainwater that are found around active volcanoes.

RNA world

Biologists have long been unhappy with aspects of the Oparin–Haldane model. They have pointed out, for example, that the two fundamental functions of any living thing are that it must have some form of genetic code, the ability to pass on information from one generation to the next, and it must be able to perform chemical reactions, to break down food, for example. These are, respectively, the functions of genes and enzymes. *Genes* are the segments of the genetic code, written in the sequence of bases in the DNA (deoxyribose nucleic acid), that specify particular functions. *Enzymes* are chemicals that stimulate, or catalyse, chemical reactions. The conundrum was to determine whether life originated according to a 'genes first' or 'enzymes first' model.

The solution seems to be that perhaps both functions arose at the same time. In 1968, Francis Crick (1916–2004) suggested that RNA was the first genetic molecule. He argued that RNA could have the unique property of acting both as a gene and an enzyme, so RNA on its own could be a precursor of life. RNA (ribonucleic acid) is one of the nucleic acids and it has key roles in *protein synthesis* within the cells. The *genetic code*, the basic instructions that contain all the information to construct a living organism, is encoded in the DNA strands that make up the chromosomes. Different forms of RNA act as the template for translation of genes into proteins, transfer amino acids to the *ribosome* (the cell organelle where protein synthesis takes place) to form proteins, and also translate the transcript into proteins.

When Walter Gilbert from Harvard University first used the term 'RNA world' in 1986, the concept was controversial. But the first evidence came soon after when Sidney Altman of Yale University and Thomas Cech of the University of Colorado independently discovered a kind of RNA that could edit out unnecessary parts of the message it carried before delivering it to the ribosome. Because RNA was acting like an enzyme, Cech called his discovery a *ribozyme*. This was such a major finding that the two were awarded the Nobel Prize for Chemistry in 1989; Altman and Cech had confirmed part of Crick's prediction.

But how could naked RNA molecules exist, and how could they act as a foundation for life? The argument was that the simple RNA molecules may have assembled themselves by chance in rock pools, more or less following the assumptions made by Oparin and Haldane, and as shown in the Stanley Miller experiment. These simple naked RNA molecules mainly existed and then disappeared, but perhaps one or two were able to copy themselves, and they could have become dominant.

To take this forward to create a living cell, there might have been two stages, the production of a protocell by combination of two components, an RNA enzyme and a self-replicating vesicle (Fig. 4). This satisfies the minimum requirement that two RNA molecules should interact, one to act as the enzyme to bring together the components, and the other to act as the gene/template. Together the template and the enzyme RNA combine as an *RNA replicase*. But these components have to be kept together inside some form of compartment or cell, or they would only occasionally come into contact to work together. This is the second pre-life structure, termed a *self-replicating vesicle*, a membrane-bound structure composed mainly of *lipids* (organic compounds that are not soluble in water, including fats) that grows and divides from time to time. The RNA replicase at some point entered a self-replicating vesicle, and this allowed the RNA replicase to function efficiently (Fig. 4).

This is a protocell, but it is not yet living. It is just a self-replicating membrane bag with an independent self-replicating molecule inside. To make the protocell function both components have to interact, the vesicle protecting the RNA replicase, and the RNA replicase perhaps producing lipids for the vesicle. If the interaction works, the protocell has become a living cell. The cell is alive because it has the ability to feed itself, to grow, and to replicate. Evolution can happen because the cells show differential survival ('survival of the fittest'), and the genetic information for replication is coded in the RNA.

Some aspects of the RNA world hypothesis have been tested, but much remains to be done. And in any case, the model remains hypothetical, because none of these stages would be likely to be fossilized. If the RNA world existed, it had to pre-date the oldest fossils, and the Earth had to be cool enough for the organic elements to survive being burned off. Some estimate that this might have been a time of 100–400 million years, somewhere between 4.0 and 3.5 billion years ago.

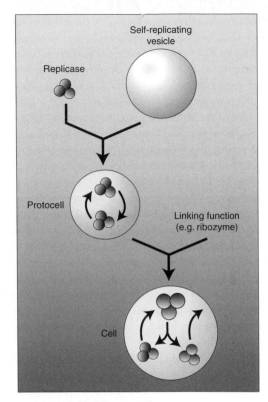

4. The formation of an RNA protocell

The first fossils

The oldest fossils appear to date from about 3.5 million years ago. Fossils of this age have always been controversial, but there are two kinds, microfossils and stromatolites. The first truly ancient fossils were reported in the 1950s, and the pressure to find ever-older specimens is intense. Mistakes have often been made, and that is no surprise because the oldest fossils are bound to be from extremely simple organisms, and microscopic ones at that. So it's no wonder that great experts have often been caught out over-interpreting a chance bubble or mineral fragment in a microscope slide, even a bit of fluff or a modern plant spore.

It is probably unexpected that the most convincing truly ancient fossils are large structures called *stromatolites*. These are mounds made partly from living organisms and partly from sediment, and they still exist today. Stromatolites (Fig. 5a) are made from many thin layers that apparently build up over many years or hundreds of years to form irregular mushroom- or cabbage-shaped structures. They are built from microbial mats composed of some of the simplest of living organisms called *cyanobacteria*, and these have sometimes been called, rather misleadingly, blue-green algae. Algae, like seaweeds, have advanced cells with nuclei, whereas cyanobacteria, like ordinary bacteria, are made from the simplest of cells, without a nucleus.

Typical cyanobacteria photosynthesize, so they live in shallow water, near the water's edge. Today, they are found generally in highly saline waters, often in tropical regions, where pools of seawater have partly evaporated. In less saline waters, herbivorous animals eat them up. The thin microbial mat may sometimes then be swamped by fine grains of mud, and the cyanobacteria grow up through the sediment to keep in touch with the sunlight. Over time, extensive layered structures may build up. In most fossil examples, the constructing microbes are not preserved, but the layered structure remains. Many early examples have proved controversial, but the oldest that are generally accepted come from Australia, and are dated as 3.43 billion years old.

Perhaps the oldest currently accepted microfossils other than stromatolites date from 3.2 billion years ago. They were reported in 2000, from a massive sulphide deposit in Western Australia. The fossils are thread-like filaments (Fig. 5b) that may be straight, sinuous, or sharply curved, and even tightly intertwined in some areas. The overall shape, uniform width, and lack of orientation all tend to confirm that these might really be fossils, and not merely inorganic structures. If so, they confirm that some of the earliest life may have been *thermophilic* ('heat-loving') bacteria that lived near a hot, sulphur-producing structure under the sea, as predicted by Euan Nisbet and Norman Sleep's model for the origin of life.

There is a long gap in time after the 3.4-billion-year-old stromatolites and microfossils before more convincing fossils are found. There are some specimens from rocks dated at 2.5 billion years old in South Africa, and then the famous Gunflint Chert of Canada, dated at 1.9 billion years ago. The Gunflint microfossils include six distinctive forms, some shaped like filaments, others spherical, and some branched, or bearing an umbrella-like structure. These Precambrian cells resemble in shape various modern bacteria, and some were found within stromatolites. Most unusual is *Kakabekia*, the umbrella-shaped microfossil; it is most like rare micro-organisms found today at the foot of the walls of Harlech Castle in Wales. These modern forms are tolerant of ammonia (NH_3), produced by ancient Britons urinating against

5a. Stromatolite fossils in the Stark Formation, Mackenzie, Canada

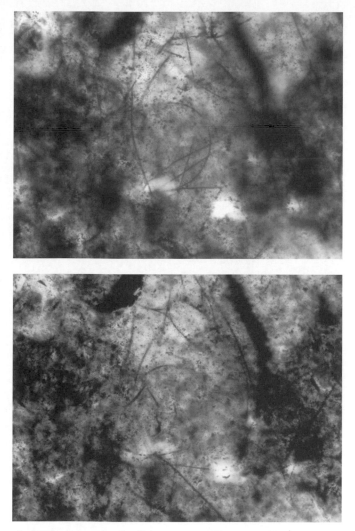

5b. Filamentous microfossils in a 3,235-million-year-old massive sulfide from Australia

the castle walls. So were conditions in Gunflint Chert times also rich in ammonia?

Strange things were happening on the Earth 2 billion years ago, apart from the ammonia-loving *Kakabekia*. The atmosphere suddenly seemed to carry oxygen, there are organic traces of quite diverse life, and new kinds of microfossils appear, some of them with nuclei. If this is true, these mark the origin of the eukaryotes, and so the origin of sex.

Chapter 2
The origin of sex

What use is sex?

John Maynard Smith, 'The origin and maintenance of sex' (1971)

It has often been noted that sex is a ludicrous and messy business. Simple organisms seem to be able to reproduce perfectly successfully by splitting or budding: amoebas go on feeding until they are quite large, and then one individual splits into two; a yeast or a sponge buds off side shoots that eventually break free as separate little organisms. So what's the point of sex?

In his book *The evolution of sex*, the noted British evolutionary thinker John Maynard Smith (1920–2004) wrote in 1978 about the twofold cost of sex. He pointed out that *asexual* organisms, those that have only one gender and that reproduce by splitting or budding, can increase their population sizes rapidly. Because each individual is effectively a female, each of the offspring is capable of reproducing independently. *Sexual* organisms, those that reproduce following exchange of genetic material, have two sexes, female and male, and it's the males (of course) that are the problem. So if each female produces two offspring, and there is 1 : 1 sex ratio, then on average the two offspring will consist of one female and one male. The rate of doubling of the population size is half that of an equivalent asexual organism.

Technically speaking, the sexual female has half the *fitness* of the asexual female. Fitness, genetically speaking, is a measure of reproductive success. So the 'twofold cost of sex' is that a sexual organism has half the fitness of its asexual counterpart.

So what is it about sex that has made it such a worthwhile pursuit? Maynard Smith suggested that the advantage was a long-term one, that sex shuffles genes more effectively than *parthenogenesis* (the production of live young from unfertilized eggs), introducing more genetic variability, and hence adaptability, into a population. He showed that sexual populations can evolve more rapidly than asexual ones, an ability that makes species which reproduce sexually much more resilient when the population is attacked by disease or parasites. The balance of advantage can go both ways. Normally asexual organisms such as aphids may pass through occasional sexual generations. Equally, parthenogenesis has evolved many times among lizards and snakes, groups that are typically sexual, of course.

Sex requires the transfer of genetic material between the male and female, and it is a feature unique to eukaryotes, the more complex organisms. So when, in the rather obscure history of Precambrian life, did eukaryotes arise, and then when did sex first happen? The evidence comes partly from the study of modern organisms, partly from geochemical studies of biomarkers, partly from investigations of ancient atmospheres, and partly from fossil specimens.

The universal tree of life

In the popular mind, and probably in many older biology textbooks, all of life can be divided comfortably into plants, animals, and microbes. Plants are green and they don't move, animals are usually not green and they usually move, and microbes are just small.

This rather unsophisticated classification has been supplemented and revised substantially. First, there is clearly a deep division between the *prokaryotes* and the *eukaryotes*. Prokaryotes are all single cells, they have no nucleus, and they have merely a single strand of DNA that carries all their genetic material. They generally reproduce asexually, although many forms have processes for sharing genetic material. Eukaryotes include many single-celled forms, but also many multicelled plants and animals also. Their cells include *organelles*, specialized structures such as the nucleus, energy-transmitting structures called mitochondria, and photosynthesizing chloroplasts in green plants. Their DNA is typically in many strands, forming chromosomes within the nucleus of each cell.

A five-kingdom classification of life was popular for a while, with plants and animals supplemented by fungi among the larger forms, and two major groups of microscopic organisms, the eukaryotic protoctists and the prokaryotic monerans. The five-kingdom model was demolished after 1977 in a remarkable series of papers by Carl Woese and colleagues from the University of Illinois. Their molecular trees showed a deep split into three fundamental divisions, the domains Bacteria (or Eubacteria), Archaea (or Archaebacteria), and Eucarya (or Eukaryota). So the prokaryotes are no more, forming the domains Bacteria and Archaea, and it is still not clear whether Archaea and Bacteria split first, or Archaea and Eucarya. Despite this uncertainty at the root, Woese had produced the first universal tree of life (Fig. 6).

So, all living things fall into these three great domains. The Domain Bacteria includes Cyanobacteria and most groups commonly called bacteria. The Domain Archaea ('ancient ones') comprises the Halobacteria (salt-digesters), Methanobacteria (methane-producers), Eocytes (heat-loving sulphur-metabolizing bacteria), and others. The Domain Eucarya includes an array of single-celled forms that are often lumped together as 'algae', as

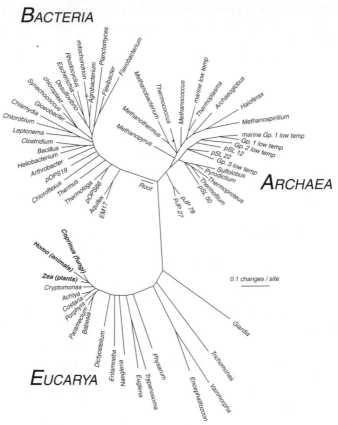

6. The universal tree of life

well as the multicellular organisms. Perhaps the most startling observation is that, within Eucarya, the fungi are more closely related to the animals than to the plants, and this has been confirmed in several analyses. This poses a moral dilemma for vegetarians: should they eat mushrooms or not?

The origin of eukaryotes

Until recently, it seemed clear that prokaryotes had dominated the Earth for a billion years or more, before the first eukaryotes appeared. However the evidence is far from clear now. First, as we have seen, molecular reconstructions of the universal tree of life do not confirm that Eucarya arose later than Bacteria or Archaea, as had been expected. In fact, all three domains might have arisen at about the same time for all we know. Geochemical data from biomarkers has also given surprising evidence.

Biomarkers are organic chemical indicators of life. Most biomarkers are *lipids*, fatty and waxy compounds found in living cells. Some biomarkers are indicative of life in general, but others can be associated with particular domains or kingdoms. In 1999, Jochen Brocks, a research fellow at Harvard, and colleagues, announced new biomarker evidence from organic-rich shales in Australia dated at 2.7 billion years ago. As expected, some of the biomarkers were indicators of cyanobacteria, but the investigators also unexpectedly identified C28–C30 steranes, which are sedimentary molecules derived from sterols. Such large-ring sterols are synthesized only by eukaryotes, and not by prokaryotes. So, this biomarker evidence confirms the existence of cyanobacteria at least 2.7 billion years ago, but it is also the oldest hint of the occurrence of eukaryotes, long before any fossils.

But how could eukaryotes, with their complex internal structure of nucleus and other organelles, have arisen from simpler prokaryotes? The most popular idea has been the *endosymbiotic theory*, proposed by Lynn Margulis, then a young faculty member at Boston University, in 1967. According to her theory (Fig. 7), a prokaryote consumed, or was invaded by, some smaller energy-producing prokaryotes, and the two species evolved to live together in a mutually beneficial way. The small invader was protected by its large host, and the larger organism received

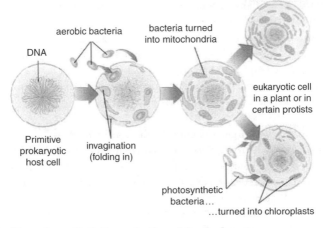

7. The endosymbiotic theory for the origin of eukaryotes

supplies of sugars. These invaders became the mitochondria of modern eukaryote cells. Other invaders may have included worm-like swimming prokaryotes (spirochaetes) that became motile flagella (the whip-like appendages used by some micro-organisms to get around), and photosynthesizing prokaryotes that became the chloroplasts of plants.

The endosymbiotic model is immensely attractive, and some aspects have been confirmed spectacularly. Most notable is that the mitochondria and chloroplasts in modern eukaryotes are confirmed as prokaryotes, the mitochondria being closely related to alpha-proteobacteria and the chloroplasts to cyanobacteria. So the amazing thing is that a modern eukaryote cell has proven prokaryotic invaders that possess their own DNA and that coordinate their cell divisions with the divisions of the larger host cell.

Many experts reject the endosymbiotic theory, or at least most of it. They point out that the only real evidence for engulfment is for the mitochondria. There is no evidence to support the idea that the nucleus was engulfed, nor is it clear what kind of prokaryote did the engulfing, and in fact engulfment is seen today only among eukaryotes, and not among prokaryotes. So the alternative view, termed the *protoeukaryotic host theory*, is that an ancestral eukaryote, the so-called protoeukaryote, already equipped with a nucleus, indeed did engulf an energy-transferring prokaryote that became the mitochondrion. But this does not tell us where the protoeukaryote itself came from. Further doubt is cast on the classic endosymbiotic theory by the suggestion that neither Archaea nor Bacteria appear to be ancestral to Eucarya, and that biomarker evidence indicates an unexpectedly ancient origin for eukaryotes. Back to the drawing board!

Oxygen

The atmosphere of the earliest Earth was devoid of oxygen, and life originated in the absence of oxygen. Then, about 2.4 billion years ago, perhaps a billion years after life had first appeared, atmospheric oxygen levels rose to 1 or 2 per cent of modern levels. This may not sound much, but geologists have termed this grandly the Great Oxygenation Event. The world would never be the same again. But what caused this rather dramatic change in the atmosphere?

The first organisms had *anaerobic* metabolisms, that is, they operated in the absence of oxygen. Indeed the first prokaryotes would have been killed by oxygen. This is a shocking fact that is confirmed by living microbes: some can switch from anaerobic to aerobic respiration depending on oxygen levels. Others, though, are obligate anaerobes that have to respire anaerobically and cannot survive in the presence of even the smallest amount of oxygen.

The simplistic view is that organisms produced the atmospheric oxygen, and that may be partly true. Gaseous oxygen is a major product of photosynthesis, and it is likely that the earliest cyanobacteria could photosynthesize. But it is unlikely that all the early atmospheric oxygen came from photosynthesis because cyanobacteria had been around since 3.5 billion years ago (see p. 29), and they had not oxygenated the atmosphere at all during the subsequent billion years. Probably all the oxygen produced by photosynthesis was mopped up by combining with gases produced by volcanoes and with soluble metals in hot springs and seafloor vents to produce water and oxides. This left little or no oxygen to enter the earliest atmosphere as a gas.

So where did the first atmospheric oxygen come from? David Catling at the University of Bristol argues that the source, initially at least, was inorganic. He suggests that the key is in methane. Methane, a compound of carbon and hydrogen, is a potent greenhouse gas produced largely by anaerobic microbes. Before life existed in abundance, there was not much methane, but levels rose as more and more was generated by the early microbes. Today, methane is consumed by oxygen in the atmosphere, but in the absence of oxygen, early Precambrian methane levels might have been 100 to 1,500 times as high as today. This created a burning hot greenhouse climate worldwide.

The methane greenhouse collapsed 2.4 billion years ago. As methane levels rose, hydrogen atoms were transferred out of the Earth's atmosphere into space, so there was no longer enough hydrogen to combine with free oxygen to form water molecules (H_2O), and so the surplus oxygen flooded out as an atmospheric gas. The rise of oxygen in the atmosphere had a profound effect on life and the planet. New *aerobic* organisms arose that exploited the atmospheric oxygen molecules in their chemical activity. The oxygen also built up an *ozone layer* high in the atmosphere that blocks out solar ultraviolet radiation.

There was a second rise in atmospheric oxygen, to 10 per cent of modern levels, around 0.8–0.6 billion years ago, and this might indicate further changes in global chemical cycles, and further expansions in the diversity of life on Earth. Such are the indicators from the rocks.

First eukaryote fossils

At one time, there was a clear story of prokaryotes-first, eukaryotes-second. As we have seen, however, the waters are considerably muddied by new molecular and chemical evidence. There are clear-cut biomarkers for eukaryotes dating from 2.7 billion years ago, and the universal tree of life resolutely refuses to resolve itself in a clear way to show that Eucarya is a younger branch than either Bacteria or Archaea.

The fossils are equally ambiguous. Textbooks used to illustrate nice cells with clear nuclei, from the Bitter Springs Chert of Australia, dated at 800 million years ago. Some of the Bitter Springs fossils even seemed to show cell division: a sensational discovery! In the way of things, of course, these were too good to be true, and they are now reinterpreted as clusters of cyanobacteria. The supposed nuclei, dark splodges on the cell, are interpreted now as folds and irregularities in the cell membranes. Further, basic cell division, called technically *mitosis*, where one cell splits more or less equally into two, is seen in both eukaryotes and prokaryotes. As Thomas Henry Huxley once said, it is terrible to see 'the slaying of a beautiful theory by an ugly fact'; in this case, several ugly facts.

The oldest supposed eukaryote fossil is impressive in some ways, disappointing in others. Later Precambrian rocks of various ages have yielded examples of a strange fossil, consisting of great spaghetti-like coils of tubes about 5 millimetres wide, preserved as thin carbon films, which have been called *Grypania*. The oldest *Grypania* fossils date from 1.85 billion years ago. The fossil looks most like a coiled seaweed and, if that's what it was, then it is a eukaryote. This interpretation is disputed, and other researchers say it is some kind of giant bacterium. They claim that the oldest eukaryote fossils are actually microscopic fossils called acritarchs, marine plant-like planktonic organisms that were roughly spherical and carried tiny whiskers and hooks. The oldest acritarchs are 1.45 billion years old.

These early supposed eukaryotes are still disputed, but by about 1 billion years ago, there are several contenders, both microscopic acritarchs and relatives, as well as numerous seaweeds, and other rather more complex fossils. The current view is that multicellularity and sex may be linked.

Multicellularity . . .

There are no truly multicellular organisms among the Bacteria or Archaea. Admittedly, some prokaryotes form filaments and loose aggregations of cells, but these associated cells do not exchange messages and their functions are not coordinated. So all truly multicellular organisms are eukaryotes, so far as we know. The simplest multicellular organisms are microscopic, and consist of little more than a string of identical cells, but some modern algae give clues about how multicellularity might have arisen.

The slime mould *Dictyostelium* generally operates as a single cell, but at certain times, notably when food is limited, numerous individual cells aggregate together and the whole colony moves to a new location. Other simple eukaryotes today, such as the protozoan *Volvox*, can form colonies of up to 10,000 individual cells, and these may show some cell differentiation. *Volvox* has fascinated scientists for years; when Anton van Leeuwenhoek (1632–1723), the famous inventor of the first microscope, first viewed *Volvox* he could not believe what he saw. The colony formed a hollow ball, and moved through the water seemingly by rolling (the name *Volvox* means 'fierce roller'). Most of the 10,000 cells act as feeding and swimming organs, beating furiously with their flagella, and causing the whole colony to spin. But small numbers of cells in the colony can take on a reproductive function, and *Volvox* colonies/individuals can mate and produce dormant offspring. In nature, *Volvox* reproduces asexually, and the sexual offspring seem to be an insurance against particularly bad conditions.

This example illustrates all kinds of extraordinary biological principles. First, where do you draw the line between an individual and a colony? The *Volvox* ball seems to act as an individual, in that the cells all stick together and work together to make it swim. But each cell is still essentially an individual, acting on its own to feed and split from time to time. Other examples of colonies today are found on coral reefs, where numerous individual corals of one species grow together as a single structure, or an ant's nest, where numerous specialized kinds of ants work together. The individual components of the colony (the coral, the ant) can live on their own and perhaps found a new colony, although that is not really true for most individual colonial ants – they rely on others to reproduce, find food, protect the nest, or keep the nest cool.

But what are the advantages of multicellularity? These must be many, because multicellularity has arisen independently many times, and it is still evolving in certain algae such as *Volvox*. Advantages of multicellularity include greater efficiency in feeding, movement, reproduction, and defence by having specialized cells. A specialized cell that only has to feed or provide a virulent defensive capability can perhaps evolve much further and specialize to a much greater extent than a single cell ever could if it has to provide all the normal services and functions of life. There are also clearly advantages in being larger than microscopic, not necessarily because big is always best, but if you are the only large organism in a sea of midgets. These advantages include access to new food sources, including larger prey, and the possibility of moving faster and further.

. . . and sex

But where does sex come in? After all, some prokaryotes and single-celled eukaryotes reproduce sexually from time to time. But Nick Butterfield from the University of Cambridge argues that true sexual reproduction enabled multicellularity to arise, and the two appear to be intimately linked. Asexual reproduction, or budding as it is sometimes called, is really just a form of growth: cells feed and grow in size, and when they are big enough they split by mitosis to form two organisms. The DNA splits at the same time and is shared by the two new cells. The products of asexual reproduction are *clones*, being genetically identical replicas.

Sexual reproduction, on the other hand, involves the exchange of gametes (sperms and eggs) between organisms. Typically, the male provides sperm that fertilize the egg from the female. Gametes have half the normal DNA complement, and the two half DNA sets zip together to produce a different genome in the offspring, but clearly sharing features of father and mother. In eukaryotes, the DNA exists as two copies, each strand forming one half of the double helix structure. Cell divisions in sexual reproduction are called *meiosis*, where the DNA unzips to form two single copies, one going into each gamete, prior to fusion after fertilization.

Butterfield's argument is that the advantages of multicellularity are so clear that this property would have arisen as soon as sexual reproduction had appeared. No asexual organism can evolve true multicellularity because asexual organisms do not evolve in the normal way. As clones, there is little opportunity for change and for natural selection. Evolution is possible, of course, but there is no speciation, the formation of new species, in the sense we see among multicellular animals and plants.

How do we date the origin of sex? Butterfield argues that there are two lines of evidence, one phylogenetic, and one based on the tight link of sex and multicellularity. The phylogenetic argument is based on the tree of life. If we can draw a tree of relationships, we can then map certain characters onto the tree on the basis of living organisms, and then track them down to the root. We must be certain that the characters in question are true *homologues*, that is, features that arose once only and are not convergences. The argument is that sexual reproduction, as seen in modern eukaryotes, is so complex that it arose only once, so the point of origin of sex can be marked on the tree of eukaryote evolution near the base.

The other argument is based on fossils. Find a multicellular fossil, says Butterfield, and you have found sex. At present, the oldest accepted multicellular eukaryote fossil is an extraordinary organism called *Bangiomorpha* from the Hunting Formation of Canada, dated as 1.2 billion years old.

Bangiomorpha: what's in a name?

Red algae (rhodophytes) are relatively common forms of seaweed today, seen on shorelines around the world, and forming a staple part of some cuisines, such as Japanese nori. Red algae range from single cells to large ornate structures, and they may be tolerant of a wide variety of conditions. The modern red alga *Bangia*, for example, can survive in a full range of salinities, from the sea to freshwater lakes. The oldest red alga was reported in 1990, and named *Bangiomorpha* because it resembled the modern *Bangia* in certain ways, but also perhaps for other reasons.

When he named *Bangiomorpha* in 2000, Nick Butterfield employed all the smutty medieval humour of England in explaining why he had chosen the name. Its full name is *Bangiomorpha pubescens*, the species name *pubescens* chosen 'with reference to its pubescent or hairlike form, as well as the connotations of having achieved sexual maturity'. The name *Bangiomorpha pubescens* has even made it into the dictionaries of

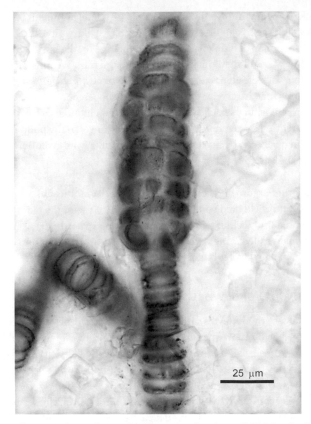

8. A close-up of *Bangiomorpha* filaments, showing cell division in the terminal structure

bizarre and cheeky names; one website notes: 'The fossil shows the first recorded sex act, 1.2 billion years ago. The "bang" in the name was intended as a euphemism for sex.' The fossils do not show sex acts, and the commentators surely exaggerate, but the name is a useful mnemonic.

Bangiomorpha grew in tufts of whiskery strands attached to shoreline rocks by holdfast structures made from several cells

(Fig. 8). The individual filaments are up to 2 millimetres long, and the cells are less than 50 microns (thousandths of a millimetre) wide. The cell walls are dark and enclose circular to disc-like cells, and filaments may be composed of a single series of cells, or of several series running side by side.

Many dozens of specimens of *Bangiomorpha* have been found, and these show how the filaments developed. Starting with a single cell, the filament grew by division of cells (mitosis) along the filament axis. One cell divided into two, then two into four, and so on. Along the filaments, disc-shaped cells occur in clusters of two, four, or eight, and these reflect further cell divisions within the filament. Some broader filaments show clusters of spherical spore-like structures at the top end; if correctly identified, these prove that sexual reproduction and meiosis were taking place. Close study of the filaments, and of series of developmental stages, shows that *Bangiomorpha* was not only multicellular, but it showed differentiation of cells (holdfast cells vs. filament cells), multiple cycles of cell division, differentiated spores, and sexually differentiated whole plants.

The Neoproterozoic and Snowball Earth

The last phase of Precambrian time is called the Neoproterozoic, a term applied to rocks dated from 1,000 to 542 million years ago. During this time, the diversity of fossils increases. This might reflect a real burst of new life forms following the invention of sex and multicellularity, or it might simply reflect the fact that it is perhaps easier for palaeontologists to find larger fossils that are visible to the naked eye. Some quite remarkable multicelled animals appeared about 575 to 565 million years ago.

The world was also changing rapidly. Oxygen had appeared and then increased in the atmosphere in two bursts, as we have seen. The Earth might also have gone through a period of freezing, called the Cryogenian, but more graphically termed 'Snowball Earth'.

The concept of Snowball Earth is highly controversial. There is no doubt that much of the Earth was cold for a long time in the Neoproterozoic: geologists had long noted evidence for glaciation such as glacial tills (rocks ground to dust by glaciers), scratches produced by the passage of glaciers carrying boulders, and dropstones, rocks dropped from the bases of icebergs into marine sediments. For many geologists, this simply showed that there had been ice caps at the Neoproterozoic poles, but for others it meant something quite different.

Joseph Kirschvink, a professor at the California Institute of Technology, coined the term 'Snowball Earth' in 1992, and envisaged a world that was almost completely covered with snow from the poles to the equator. He invoked the evidence of glacial sediments, including some examples from regions that apparently lay near the Neoproterozoic equator, and his work was extended and promoted by Paul Hoffman, from Harvard University, based on his studies of Neoproterozoic successions in Namibia.

Hoffman and others have presented extensive evidence from Neoproterozoic sediments that the Earth was entirely icebound for millions of years, and then the ice melted during a subsequent greenhouse phase as a result of massive volcanic eruptions with the production of copious amounts of carbon dioxide. Advocates of the Snowball Earth suggest that life survived under the ice, and did not diversify greatly until melting ensued. Critics suggest that it is impossible for the Earth to freeze over completely, and that at least there must have been habitable oceans around the equator. Whether the Earth was entirely or largely covered in ice, there certainly were major glacial episodes in the Neoproterozoic, and complex multicellular organisms appeared only after the glacial episodes had ended. These were the Ediacara faunas.

The fossils of the Ediacara Hills

Palaeontologists had occasionally found strange frond-like structures in rather ancient sandstones, perhaps Precambrian, perhaps Cambrian in age, but they had been unable to interpret them. One such finding happened in 1946, when Reginald Sprigg, a young mining geologist, was prospecting through the Ediacara Hills, north of Adelaide, Australia. He found round impressions that looked like jellyfish, branching fronds, and worm-like impressions.

When Sprigg reported his findings, the Ediacara Hills became famous, and the particular fossil assemblage has been called Ediacaran; this is also the name for the time interval marking the last part of the Neoproterozoic. Ediacaran organisms have been reported from more than thirty localities, from Australia, Africa, Europe, and elsewhere. The Ediacaran fossils are mostly about the same age, some 575 to 542 million years old, and they are the first true fauna, that is, life assemblage, of diverse complex organisms on Earth.

9. Life as it may have looked in Ediacaran times

More than a hundred species of Ediacaran animals have been named (Fig. 9). Most of them have been classified in modern groups, such as jellyfish, worms, and sea pens, but this is very hard to confirm. Others have argued that the Ediacaran animals represent a completely independent radiation of organisms that do not link with later, Cambrian, faunas. One researcher has identified all Ediacaran organisms as fungi, whereas Dolf Seilacher from the University of Tübingen has argued that they are unique structures that represent an independent diversification of animals that resolved structural problems in ways quite unlike anything now living. He argued that the skin must have been flexible, although it could crease and fracture, and it must have allowed oxygen and waste materials to diffuse in and out. The vendobionts, as he termed them, were interpreted as unique pneumatic structures, like car tyres or blow-up mattresses. Their outer surfaces enclosed a gas-filled interior, and their radial and segmented structures are like the divisions of a modern bouncy castle or air mattress, designed to maintain strength and flexibility.

Whatever they were, whether early jellyfishes and worms, or proto-bouncy castles, the Ediacaran faunas worldwide died out about 540 million years ago. But their demise did not leave the Earth devoid of life. Indeed, one of the greatest events in the history of life was about to happen, the Cambrian Explosion.

Chapter 3
The origin of skeletons

The fossil record had caused Darwin more grief than joy. Nothing distressed him more than the Cambrian explosion, the coincident appearance of almost all complex organic designs.

Stephen Jay Gould, *The Panda's Thumb* (1980)

The appearance of skeletons in the fossil record some 540 million years ago has long been a puzzle. It is not perhaps such a puzzle that scientists throw in the towel, as creationist critics gleefully report on their websites, but a real problem to be resolved. The fact is that, shortly after the beginning of the Cambrian period, currently dated at 542 million years ago, and some time after the extinction of the Ediacaran organisms, a broad diversity of animals with skeletons appeared in the sea. A *skeleton* to a biologist is any kind of mineralized, or partly mineralized, structure that acts as a support or framework for an organism. So our internal skeleton of bones fits the bill, but so too do the calcareous shells of molluscs and corals, the outer cuticles of insects and crabs, and even arguably the woody stems of trees.

The Ediacaran fossils of the Neoproterozoic did not have shells or skeletons of any kind we would recognize today. Perhaps, as Dolf Seilacher suggests, they had a quilted pneumatic structure that stiffened their bodies and allowed them to reach reasonable body size. Then, in Lower Cambrian rocks around the world, a diversity of shelly fossils appears. It is the fact that skeletonized organisms seem to appear suddenly, geologically speaking, and all at the same time, that is the puzzle. Why, for example, don't we first find sponges with skeletons of spicules, then corals with their tube-like houses, then perhaps shellfish with their encapsulating valves, and so on? Of course, when looking back over half a billion years, it's not easy to date every rock formation precisely, but every study seems to suggest a rather coordinated appearance of animals with skeletons about 542 million years ago. This dramatic event has been called the *Cambrian Explosion*.

The debate revolves around the reality of this event. Most palaeontologists and evolutionists, including Darwin, have suggested that the Cambrian Explosion was real and that what you see actually happened. Others, however, urge caution and suggest that we might be seeing something artificial, the result perhaps of incomplete preservation of the fossils. It could be, for example, that there are major gaps in the rock record at the end of the Neoproterozoic, or that the sediments that were deposited through that interval were not the right ones to preserve mineralized skeletons. In this chapter we will explore what skeletons are, what the fossil and rock record shows, new molecular evidence, and the rather heated debates about whether the Cambrian Explosion is real or not.

Skeletons

Skeletons are not just for physical support, although that is a major, often *the* major, function. They also provide sites for the attachment of muscles and a mineral store. So, for example, in humans, we rely on the framework of our skeleton to be able to walk and eat. The muscles attach at both ends to bones in the skeleton, and muscle contractions make the arms and legs work. In feeding, jaw muscles pull the lower jaw up and down against the skull, and the jawbones carry the teeth, all essential in feeding.

Bone is composed of two main components, the protein collagen and spicules of apatite, a form of calcium phosphate. Collagen is the primary component of cartilage. We have cartilage in our noses and ears, and it is a bendy kind of unmineralized bone. Among living vertebrates, the backboned animals, sharks have almost entirely cartilaginous skeletons that only occasionally become mineralized (and of course their teeth are mineralized), and it seems that the Cambrian predecessors of modern fishes also mostly had cartilage skeletons.

Our bones also act as mineral stores. When we are young and growing, the body has to scavenge large amounts of calcium and phosphorus from our food and it passes through the blood vessels to the bones. If a person is starved at a young age, their bones cannot grow properly, and they become stunted. Later in life, calcium and phosphorus may be mobilized from within the bones when they are needed. Bone is living, laced through with blood vessels, and other tissues. If food is short, calcium and phosphorus are absorbed from the bone back into the blood supply and passed to the cells where it is needed. The minerals can be replaced later when food is abundant. So if you were to cut through any of your bones, you would see evidence for how it grew to its present size during your childhood. You would also see evidence for episodic extraction and replacement of calcium and phosphorus in the form of channels that are widened as minerals are extracted, and that fill up in layers as minerals are replaced, rather like a water pipe furring up in an area of hard water.

Other animals have different kinds of skeletons. Skeletons may be composed from inorganic mineralized materials, such as forms of calcium carbonate, silica, phosphates, and iron oxides. Calcium carbonate makes up the shells of microscopic foraminifera, some sponges, corals, bryozoans (colonial creatures), brachiopods ('lamp shells'), molluscs, many arthropods (trilobites, crabs, insects), and echinoderms (sea urchins, sea lilies). Silica forms the skeletons of radiolarians (planktonic organisms) and most sponges, while phosphate, usually in the form of apatite, is typical of vertebrate bone, as we have seen, and the shells of certain brachiopods and the tiny toothed jaw structures of certain worms. There are also organic hard tissues, such as lignin, cellulose, sporopollenin, and others in plants, and chitin, collagen, and keratin in animals, which may exist in isolation or in association with mineralized tissues.

The simplest skeletons are seen in the sponges, which are composed of loose aggregates of spicules, pointed microscopic structures made from calcium carbonate or silica. Most other animals have an external skeleton, or *exoskeleton*. (Humans, and other vertebrates, have an internal skeleton, or *endoskeleton*.) In corals, brachiopods, and molluscs, the exoskeleton is a layered structure, built up year by year, or month by month, with growth lines often visible on the outer surface and in cross-sections. Other animals shed their exoskeletons – animals such as arthropods, nematode worms, and some rarer groups. Indeed, skeleton-shedding may be a unique feature of this particular group.

The diversity of skeleton types, and the fact they are constructed in so many different ways – some are internal, some external, some are shed, and others are not, they may be made of different mineral constituents – makes it hard to understand how skeletons seemingly evolved at the same time in all these animal groups, and everywhere in the world. What does the fossil record show us, if we follow it step by step through the transition from the latest Precambrian into the Cambrian?

Small Shelly Fauna

The first step is represented by the time of the 'Small Shelly Fauna', so called, perhaps not surprisingly, because it is a fauna that is composed of small shells. The term 'small shells', however, hides a great deal of ignorance: small shells they may be, but the affinities of many of them are unclear.

The Small Shelly Fauna (SSF) has been identified in the latest Precambrian, but is best known in Lower Cambrian rocks, dating from perhaps 542 to 530 million years ago. The importance of the SSF is that it comes before the appearance of larger fossils with skeletons, and so marks the first phase of the Cambrian Explosion.

It has proved very hard to understand the biology of the SSF animals, and they are generally named simply according to their

A

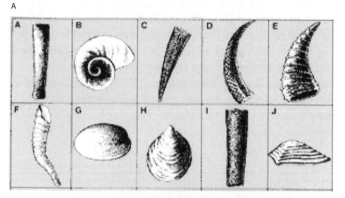

B

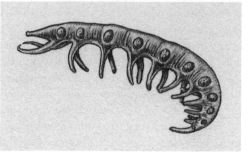

10. Fossils from the Early Cambrian. A: A selection of Small Shelly Fossils from the Siberian Precambrian-Cambrian boundary strata; B: *Microdictyon*

shapes (Fig. 10A). Two major groups are the hyolithelminthids with phosphatic tubes, open at both ends, and the tomotiids with phosphatic cone-shaped shells, usually occurring in pairs. Other animals were tube-builders that secreted carbonate walls, organic-walled tubes possibly of an unsegmented worm, and phosphatic plates, or sclerites, from larger but unknown animals.

The sclerites give clues to a whole array of animals we barely understand. Mostly their bodies have gone, and all we have are the minute, microscopic leaf-shaped sclerites. It is assumed that these fitted together as some kind of flexible armour over animals that may have looked roughly like pine cones. Some exceptionally preserved specimens from China, called *Microdictyon*, suggest that some of the sclerite-bearers at least were worm-like animals (Fig. 10B), which carried oval plates arranged in pairs along the length of the body which may have provided a base for muscle attachment associated with locomotion. What is intriguing is that some of the sclerites might have come from quite large animals that are otherwise entirely unknown, and may never be known other than by these intriguing exuviae.

The Cambrian Explosion

The Small Shelly Fauna of the Early Cambrian was a precursor of the Cambrian Explosion proper. Towards the end of the Early Cambrian, and overlapping in time with the SSF, a dozen or more major animal groups appeared. At one time, it was thought that they all appeared at once, but more careful study suggests a rather more orderly procession, with one group appearing after another. Some of the evidence comes from fossils of the organisms themselves, and other steps along the way are indicated at present only by trace fossils, tracks, and trails. This may seem rather uncertain evidence, but many tracks and trails can be quite diagnostic of their makers, especially if they show foot or leg marks, for example.

So, in sequence, the first evidence for the radiation of animals in the sea, and the first step of the Cambrian Explosion, is represented by tracks and trails dating from 555 million years ago, at the end of the Neoproterozoic. These tracks were made by elongate bilaterally symmetrical animals, mostly worms of one sort or another. Then, the trace fossil record shows the first evidence of arthropods, the 'jointed limbed' animals, in the earliest Cambrian, some 540 million years ago. Then the Cambrian Explosion proper began about 530 million years ago, with the appearance of the first skeletal fossils of trilobites and echinoderms, and it lasted for perhaps 10 million years, during which time global diversity burgeoned. Groups such as molluscs and brachiopods, which possibly appear in the SSF, are represented by unequivocal fossils. Sponge spicules are abundant in places.

If you had gone back to this time in the Early to Mid Cambrian, some of the new skeletonized creatures might have seemed familiar; others less so (Fig. 11). Brachiopods survive today, but they were much more important in the Palaeozoic, the time from 542–251 million years ago, than they are today. Brachiopods include a broad range of forms. Some, such as *Lingula*, have simple teardrop-shaped paired valves, and live in vertical burrows with a long, fleshy pedicle holding them in the sediment, and filter-feeding food particles from the water above. Most brachiopods lived on the seafloor, not in a burrow, and most adopted the classic shape of a Roman lamp, with two unequal valves. The two valves fitted together like halves of a nutshell, and they could open to allow passage of water in and out for feeding and respiration. The pedicle valve is larger than the brachial valve, and the fleshy pedicle emerges from a hole at the apex of the pedicle valve (this is the equivalent of the wick hole in a Roman lamp). Brachiopods dominated the seabed throughout the Palaeozoic.

Cambrian echinoderms were unusual beasts, somewhere between a sea urchin and a sea lily. Most of them were attached to the seabed by a stalk, and a bulbous body stood above, covered with a

11. The Burgess Shale scene, Middle Cambrian

calcium carbonate skeleton made from tightly matching polygonal plates. They often had tentacles of some kind, and used these to capture food from the water and pass it into their mouth, which usually lay at the top of the body, in the midst of the tentacles. These Cambrian echinoderms are very different from their modern kin, the sea urchins, sea lilies, and starfish we often find on the beach.

Trilobites were the characteristic fossils of the Palaeozoic, and their appearance marks the core of the Cambrian Explosion. As noted above, the first evidence of trilobites comes from fossilized tracks in the earliest Cambrian, and then body fossils appeared. Trilobites (Fig. 11) had, as their name suggests, three lobes – a central structure down the length of the body, and a lobe at either side. Their bodies are divided three ways, also from front to back:

a head shield called the cephalon, a body portion called the thorax, and a tail shield called the pygidium. The whole body is segmented from front to back, and each of the thoracic segments carried a leg and a feathery gill. Trilobites ran about on the seabed, engulfing small prey or ploughing through the sediment in search of food. The mouth was a kind of trapdoor under the cephalon. At the front of the head are antennae, feelers, used as in their modern relatives, the lobsters, crabs, and insects, to sense the environment ahead (trilobites lived in clouds of disturbed seabed mud, and sometimes in deep waters). Nonetheless, most trilobites had eyes, and often spectacular ones at that. Each eye consisted of numerous eye tubes, each with a lens, as in modern arthropods. Palaeontologists have dissected these eyes (the lens is a calcite crystal that survives unaltered by fossilization) and looked through them – how strange is that: to see the world as a trilobite saw it 500 million years ago?

Other Cambrian beasts included archaeocyathids, often conical-shaped seabed dwellers that formed modest reefs. Archaeocyathids were thought to be some kind of coral, but they are more probably related to sponges, but formed a more substantial calcium carbonate skeleton. In places, their reefs may be as deep as 10 metres. There were also more modern kinds of sponges, but they are often known only from their collapsed skeletons – merely heaps of spicules. There were strange conical shells called hyolithids, a common group in the Cambrian, but of uncertain affinities. Finally, our own phylum, the Chordata, of which the vertebrates are a sub-phylum, appeared in the Cambrian, with a variety of small leaf-shaped and tadpole-shaped little creatures that swam by flickering their flat-sided bodies from side to side.

Chengjiang: a window on the Cambrian Explosion

The Cambrian Explosion is documented in unexpected detail thanks to the astonishing good fortune of a special set of fossil sites in China. The Chengjiang fauna from Yunnan Province in southern China was discovered in 1912, but not studied in detail until the 1980s and 1990s. The rock layers at Chengjiang, the Maotianshan Shales, are some 50 metres thick, and they are dated from 525 to 520 million years ago, representing the second half of the Cambrian Explosion.

So far, some 185 species have been identified in the Chengjiang biota (= fauna plus flora): algae, jellyfish, sponges, priapulids, annelid-like worms, echinoderms, arthropods (including trilobites), and chordates (both the oldest fish in the world, as well as non-vertebrate chordates). Arthropods are the dominant organisms, making up 45 per cent of the fauna, with 40 per cent belonging to the other named groups, and perhaps 15 per cent of the species representing 'enigmatic' groups. These fossils are a complete puzzle: palaeontologists haven't the faintest idea what those 'enigmatic' organisms might be.

The fossils show all the skeletal features, but also soft tissues such as skin, gut traces, eye pigments, gill structures, and segmented muscles. The soft tissues are preserved as clay films and these are sometimes spectacularly colourful – reds, purples, yellows – because of the addition of variable amounts of iron oxides. But why the exquisite preservation? The sedimentary setting of the Chengjiang biota seems to have been a shallow sea. The sediments are mainly fine-grained – muds and siltstones – so there was not much wave or current activity. Animals that lived on the seabed, and those that swam above, must have died and their carcasses accumulated without disturbance. Because of seasonal temperature changes and a cessation of mixing, the bottom waters must have become anoxic at times, and this deterred scavengers, and speeded the replication of the muscles and other soft tissues by bacteria and clay minerals.

The arthropods from Chengjiang provide some clues to the origin of skeletons. The Chengjiang trilobites, like their later relatives, all had hard, readily fossilizable skeletons made from calcium carbonate. The remainder of the Chengjiang arthropods, over 90 per cent of the arthropod species, had much softer skeletons that lacked a mineralized component. These skeletons were made from the protein chitin, the main constituent of insect exoskeletons, for example. Some of the non-mineralized arthropods would be known only fleetingly from incomplete fossils but for sites of exceptional preservation such as Chengjiang. *Anomalocaris*, for example, known also from the Burgess Shale of Canada, ranged in length from 60 centimetres to a staggering 2 metres. This giant predator looked roughly like a trilobite, with many segments, and a head and tail region. It probably swam by flapping large flexible lobes along the side of its body, and snatched prey with its large curved, flexible armoured arms that bore barb-like spikes. It then stuffed its hapless prey into a circular mouth that was surrounded by the most astonishing structure that looks like a giant pineapple ring, but almost certainly was made from plates that slid over each other, and opened and closed like the diaphragm on an old-fashioned camera.

First chordates

Equally astonishing, and unexpected, are the early chordates from Chengjiang. Palaeontologists had always thought, perhaps rather presumptuously, that the Phylum Chordata, to which we and all other vertebrates belong, might have appeared rather later than the other phyla. Surely, after all, the chordates were somehow the pinnacle of animal evolution? Not so. A great diversity of basal chordates of one sort or another has rent the palaeontological world with high-profile disputes. The soft tissues are there to be seen, picked out in gaudy hues of purple, red, and yellow. Hundreds of specimens are removed from the Chengjiang localities, many by teams of farmers employed by different museums for that very purpose. Indeed there are as many as six teams of scientists in China pursuing this geological grail. The fossils are scrutinized in different institutions, and the written descriptions and manuscripts passed from hand to hand and jealously guarded until they are published. Every blob and squiggle of soft tissue is interpreted and reinterpreted – is that a gut trace or a nerve cord, a hint of liver or a fragment of lunch, a squashed brain or a nostril?

The most astonishing find was *Myllokunmingia*, named in 1999, the oldest fish, and so the oldest vertebrate. Over 500 specimens have been collected so far, and these all show a 3-centimetre long, streamlined little fish. The head is poorly defined, but a possible mouth is seen at the front end. Behind this are five or six gill pouches. Up to twenty-five double-V-shaped muscle blocks extend along most of the length of the body. Other internal organs include a heart cavity and a broad gut. There is a low back fin along the front two-thirds of the length of the body, and a side fin along the posterior two-thirds. It is thought to be a vertebrate because of the presence of a distinct head with possible sense organs (vertebrates are also called 'craniates', meaning animals with heads). As is always the case in such disputed territory, dozens of papers have been published about *Myllokunmingia* and its close relatives (perhaps there are three or four species; perhaps one), and the precise anatomy is disputed.

Other possible chordates include the vetulicolians, a whole class of ancient creatures known only from Chengjiang. Among modern chordates, vertebrates, with their backbones, are by far the dominant group, but there are other chordates, such as the sea squirts and amphioxus. Sea squirts as adults look nothing like vertebrates: they are fleshy bags that are fixed to the seabed and feed by pumping water in and out of their central cavity. But the clue to their true affinities is given by the larval form, a little tadpole-like free-swimmer that has a cartilaginous notochord, a stiffening rod along the back, the diagnostic feature of chordates. Amphioxus, sometimes called the lancelet, is a more convincing chordate as an adult because it retains the notochord and is free swimming throughout its life.

There are several species of vetulicolians from Chengjiang, and they all look like sausage balloons, knotted in the middle. The body is in two parts, with a bulbous section in front of, and behind, a flexible connection. There is a large mouth with a strengthened rim, and preserved internal structures include the guts. Both parts of the body appear to be crossed by transverse bands of tissue, possibly muscles, possibly strengthening tissues. On the mouth-bearing segment, presumably the front part of the body, are five circular structures in a row that have been interpreted as gill slits. The vetulicolians are certainly enigmatic. The series of gill slits suggests they are chordates, because this is a feature seen only in modern chordates. But this is disputed, and vetulicolians might lie outside the chordates, as relatives of the chordates and echinoderms. If they are chordates, some authors ally them with the sea squirts, whereas others see them as at the very base of the chordate tree.

The Chengjiang biota is perhaps less famous than the fossils from the Burgess Shale in Canada (Fig. 11). The Burgess Shale is younger, so it comes after the Cambrian Explosion had played out, but it has been studied in much more detail and for 100 years. Many of the amazing creatures from Chengjiang, such as *Anomalocaris*, are well known also from the Burgess Shale. Others such as the basal chordates are not so well represented. The story of the Burgess Shale has been explored many times in excellent and eloquent accounts.

Significance of the Cambrian Explosion

The Cambrian Explosion has generated much debate, some experts interpreting it as unique in the history of life, others seeing it as one of many such bursts of diversification, and some denying its existence. Let's assume it was real for the moment, and see what can be made of it.

The 'standard' view of the Cambrian Explosion is that all major animal groups in the sea diversified after they had acquired skeletons. But why did such a diversity of skeletons appear at the same time? Geologists had long speculated that perhaps the chemistry of the atmosphere and oceans changed profoundly at the end of the Neoproterozoic. Perhaps oxygen levels were too low for abundant larger animals to evolve, or the chemistry of the oceans allowed more carbonate and phosphate to pass into circulation and so become available for unprotected animals to capture and manufacture skeletons.

It is, frankly, hard to credit any of these rather simplistic ideas. Oxygen levels had been climbing through the Precambrian and it is not clear that a major threshold level was crossed just at the beginning of the Cambrian. Further, the relatively large Ediacaran animals had existed, though without skeletons, it must be said, some 50 million years earlier. Moreover, the idea that the mineralogy of the oceans changed and that this triggered the acquisition of skeletons across diverse groups all at the same time is also surely too mechanistic – as if organisms wait for a mineral to appear, and then numerous evolutionary lineages incorporate it into their bodies independently.

In detail, the fossils show a somewhat extended Cambrian Explosion, lasting at least 10 million years – a geological instant, but a long time to live through. More likely, the sequential acquisition of skeletons was part of a so-called 'arms race'. If one group evolved a skeleton, whether based on chitin or mineralized by carbonate or phosphate, others might have had to follow suit. If a prey group becomes armoured, the predators must learn to tackle the new defences or they will die out. One way to pierce armour is to have armoured appendages. Likewise, the rise of predatory forms such as trilobites and the monster *Anomalocaris* would exert a rather direct evolutionary pressure on all other organisms of the time to become armoured or die.

There has been another area of dispute concerning the Cambrian Explosion. Stephen Gould, in his book *Wonderful life*, made a strong case that the Cambrian was a unique time, that each species in the Burgess Shale assemblages was so astonishingly different from other forms that this was a time of unfettered, high-level evolution of fundamental body plans. He made a specific claim, that the diversity of arthropods, in terms of their basic body designs, was greater in the Cambrian than it has ever been since. He used this to create a new metaphor for evolution: that somehow in the Cambrian, the arthropods, and other animal groups, diversified so rampantly that they explored genetic possibilities to their maximum. Since then, evolution, he claimed, has been weeding out this amazing basal growth of the animal evolutionary tree. Half or more of the rampant Cambrian diversity has since been lost. This was evolution by explosion and pruning.

Most other workers rejected Gould's idea, and I think he did too later in life, realizing he had perhaps been carried away by his purple prose. In a sobering study ('ugly facts and beautiful hypotheses?'), Matthew Wills of the University of Bath carried out a thorough quantitative study of the disparity (shape variation) of the Burgess Shale arthropod fauna, and the modern arthropod fauna. He showed that the variations in shape among members of both assemblages were comparable. If one compares a lobster and a butterfly, a spider and a spider crab, a rhinoceros beetle and a mite, the disparity differences are as great as, or greater than, in the Cambrian, and these results can be generalized to compare the entire Cambrian ocean with the entire modern fauna, or to focus on one geographic region today to make it more comparable with the single locality of the Burgess Shale.

But did it really happen?

In 1996, palaeontologists were struck by a bombshell. New molecular evidence published by Greg Wray from Duke University and colleagues suggested that animals had diversified about 1200 Ma. This estimate pre-dates the Ediacaran animals by some 600 million years, and the Cambrian Explosion by 650 million years. This new evidence suggested that the Precambrian fossil record of animals (and presumably all other fossils) was even more deficient than had been assumed; the Cambrian Explosion shifted back deep into the Precambrian.

Wray had taken new DNA/RNA data on diverse groups of animals and attempted to discover their pattern of relationships. This was part of a major and hugely important undertaking that is far from complete to discover the true shape of the tree of metazoans (the technical term for animals). Based on centuries of study of anatomy and fossils, there was no real consensus about relationships among the major metazoan phyla: perhaps arthropods were related to annelids (worms), perhaps sponges were near the base of the tree, and perhaps echinoderms and chordates were close relatives. The other twenty major metazoan phyla were hard to place. Molecular evidence has certainly revolutionized our understanding of the shape of metazoan evolution, and has revealed patterns of relationships that had not been suspected before. Key discoveries have been the Bilateria, for all the bilaterally symmetrical animals such as chordates, (echinoderms, arthropods, and various worm-like creatures), and within Bilateria the Ecdysozoa (for all animals that shed their skeletons from time to time, the arthropods, nematodes, and six or seven other obscure groups).

But once you have a tree, it's good to tag it with dates. When did Metazoa originate, Bilateria, Ecdysozoa, and all the other fundamental groups and subgroups? Application of the molecular clock gave Wray and his colleagues a date of 1,200 million years for the origin of Metazoa, and this was confirmed by other studies soon after.

For several years, there was an active debate back and forwards. Some claimed that the fossils couldn't lie, and the molecular dates must be wrong (I was one of those argumentative types), others accepted the new dates and said the fossil record was sadly deficient, and yet a third party said that both dates were correct, more or less, and that the 600-million-year gap was a long spell of concealed evolution. The term 'phylogenetic fuse' was invented to describe such a situation, where a major group (here metazoans) diverges, and that is marked by the molecular date, and then the first fossils appear much later. The 'fuse' refers to the proposition that evolution continued, but the organisms were small and rare, and so not detected as fossils. At a later time, something caused the group to expand suddenly, and then the fossils are found.

Such a long span of cryptic evolution strains credulity. The 'phylogenetic fuse' might be a reasonable explanation for 5 to 10 million years of hidden evolution, but it is unlikely that a group could sustain itself for huge spans of time and not either diversify or go extinct.

The theoretical debate about the 'phylogenetic fuse' was brought to a rapid end by more recent studies, by Kevin Peterson from Dartmouth University and others. These used new molecular evidence to show that the estimated date of metazoan origins was really 650–600 million years ago, older than the oldest fossils, but not much older than the enigmatic Ediacaran faunas for example. The first analyses suffered from a variety of problems with the genes and the calculation methods. The key problem had been, however, that all the dates were extrapolated from known fossil dates for splits among early fishes and other vertebrates. And, unknown to the earlier analysts, the vertebrate molecular clock ticks rather more slowly than that for other metazoan phyla. Hence, extrapolating with a slow clock, but assuming a fast rate, extends the estimate far too deep; in fact this accounts for the virtual doubling of the estimate from 650–600 to 1,200 million years.

So, the Cambrian Explosion is back on track, more or less! There is still debate about whether, as seems likely, all the metazoan groups had appeared as naked forms well down in the Neoproterozoic, at the time of the Ediacaran faunas or even a little earlier. So, 50–100 million years of the early history of these groups is missing. And the problem of whether the Cambrian Explosion really represents the rapid acquisition of skeletons by all and sundry, or is in some way a preservational artefact, is still open.

The Cambrian Explosion is still wonderful and mysterious in equal measure. The new life that became established in the oceans – all the trilobites, brachiopods, echinoderms, chordates, molluscs, and others – continued to evolve and develop ever-more complex ecosystems through the Cambrian and into the subsequent Ordovician and Silurian periods. But something else was happening at this time – some life forms were already exploring the margins of the oceans and making the challenging leap onto land.

Chapter 4
The origin of life on land

> When they went ashore the animals that took up a land life carried
> with them a part of the sea in their bodies, a heritage which they
> passed on to their children and which even today links each land
> animal with its origin in the ancient sea.
>
> Rachel Carson, *The Sea Around Us* (1951)

All life came from the sea, and it's not just the animals that carry
some of that water-living heritage: plants do too. The classic story
is that plants emerged from the watery depths perhaps in the
Silurian or Devonian period, some 400 million years ago, and
these were followed soon after by insects and worms, and other
small animals that could find new places to live and feed among
the branches of the simplest land plants. These small creeping
things were then followed by the first vertebrates on land, when
some hefty fishy creature dragged itself over the waterside mud
and started to eat flies.

As ever, the reality is rather more complex, and new fossil
discoveries have pushed the origin of land life back much further
in time than had been imagined. Indeed, there may well have been
some simple microscopic photosynthesizing organisms around the
fringes of the seas and lakes even in Neoproterozoic times. But
does it really matter when life moved onto land ('conquered the
land' in the old phrase)? Perhaps we are just interested because we

are land-livers? In fact, life on land is hugely significant for two reasons.

The first reason is that life on land represents most of modern biodiversity. Whereas some 500,000 species live in the sea today, at least ten times that number live on land. The bulk of modern biodiversity consists of insects, but other terrestrial groups, like other arthropods (spiders, centipedes), as well as flowering plants are much more species-rich than anything in the sea. So life has really prospered after it moved onto land.

Second, life has changed the face of the Earth. Before there was life on land, there were no soils. The Earth's surface was barren rock, and rates of erosion were vast, more than ten times what they are today. Mountains were rocky crags, and lowland plains were dustbowls. As life moved onto land, soils developed (soil is just rock dust plus organic matter), and the soils and plants crept outwards from the water and covered more and more of the surface. But did this process really begin in the Neoproterozoic?

Precambrian mushrooms

When molecular biologists presented evidence that some plant groups existed 600 million years ago, palaeontologists were outraged. This proposal pushed the record of land plants back by 200 million years. However, there are indeed some excellent fossils of some possible lichens from rocks of that age. Lichens are *symbiotic* (mutually beneficial) associations of a fungus with a green, photosynthesizing organism, usually an alga or a cyanobacterium.

Then Precambrian lichens were reported in 2005 from the late Neoproterozoic rocks of Doushantuo in China, a remarkable source of exceptionally well-preserved fossils of extraordinary age. The specimens are so well preserved, even to cellular level, that most palaeobotanists are convinced by the new finds. It had long been suspected that cyanobacteria formed thin crusts on land, as they do in desert regions today, photosynthesizing and forming thin 'soils' in the Proterozoic. The oldest fossil soils, dated at 1.2 billion years old, were presumably generated by microbial or algal activity. The Doushantuo lichens prove that the surface of the land, at least close to the water, was already green at the end of the Precambrian, long before plants really conquered the land.

But of course, lichens are fungal associations and fungi are not plants (see p. 36). So, the molecular results that suggest a Precambrian origin for green plants are still highly disputed. True plants apparently did not move onto land until later.

Green plants on land

The land began to become green in the Ordovician, some 450 million years ago. The first land plants seem to have been bryophytes, commonly called mosses and liverworts. The oldest recorded fossil bryophytes are Ordovician in age, although interpretations are uncertain, and there is a possible Cambrian relative, *Parafunaria* from China.

There are additional hints that green plants were moderately diverse, at least in some locations, in the Ordovician. For example, Ordovician soils with root-like structures suggest that plants were already on land. Something substantial then seems to have happened in the Mid Ordovician, when the character of microfossil assemblages changed dramatically: spores appeared. *Spores* are airborne microscopic cells that are characteristic of land plants. So, although these earliest land plants have not yet been found as fossils, they must have been there because they were producing spores. But the nature of these spores has been debated: whether they really came from green plants, or might simply be the products of green algae.

In 2003, Charlie Wellman from the University of Sheffield showed that the Ordovician spores were probably produced by small bryophytes, perhaps like liverworts. He found detailed similarities in the spore walls to those of modern liverworts, and he also found clusters of spores packaged in a type of cuticle that looked overall like a liverwort spore-bearing organ.

Bryophytes today show special adaptations to life on land, such as a waterproof cuticle over their leaves and stems. Many also have *stomata*, specialized openings under the leaves used for controlling water loss. Some bryophytes have the unusual ability to dry up completely, and then to re-hydrate when rain falls, and continue as normal. It seems then that low-growing mossy plants invaded the land in the Ordovician, and that larger green plants came later.

Adapting to life on land

As humans, for whom swimming is a bit of a struggle, we would naturally think that the key challenge for a water-dweller on land is breathing. However, breathing air, as opposed to extracting oxygen from the water, was in fact the least of the problems of the earliest land animals. And for plants, the challenges are obviously rather different. They relate especially to obtaining nutrients and water, prevention of desiccation, and support.

Nutrition first. In water, a plant may absorb nutrients and water all over its surface, but on land, all such materials must be drawn from the ground, and passed round the tissues internally. Land plants typically have specialized roots that draw moisture and nutrient ions from the soil, which are passed through water-conducting systems that connect all cells. The system is driven by *transpiration*, a process powered by the evaporation of water from leaves and stems. As water passes out of aerial parts of the plant, fluids are drawn up into the water-conducting system hydrostatically.

Water loss is a second key problem for plants on land. Whereas in the water, fluids may pass freely in and out of a plant, land plants are covered with an impermeable covering, the waxy *cuticle*. Gas exchange may be controlled by specialized openings, the stomata (singular, stoma), often located on the underside of leaves, and as seen in some bryophytes. Typically, stomata open and close depending on carbon dioxide concentration, light intensity, and water stress.

The third problem of life on land is support. Water plants simply float, and the water renders them neutrally buoyant. Most land plants, even small ones, stand erect in order to maximize their uptake of sunlight for photosynthesis, and this requires some form of skeletal supporting structure. All land plants rely on a *hydrostatic skeleton*, a stiff framework supported by water in tubes, and some groups have evolved additional structural support by the deposition of the tough organic polymer *lignin* on the internal fibres and canals of the trunk.

The first vascular plants

The oldest known vascular plant is *Cooksonia* from the Mid Silurian, some 425 million years ago, of southern Ireland, a genus that survived for some 30 million years. *Cooksonia* (Fig. 12) is composed of cylindrical stems that branch in two at various points and are terminated by cap-shaped spore-bearing structures at the tip of each branch. The specimens of *Cooksonia* range from tiny Silurian examples, only a few millimetres tall, to larger Devonian forms up to 65 millimetres tall.

In lifelong studies of these extraordinary little plants, Dianne Edwards of the University of Cardiff has discovered spores in the

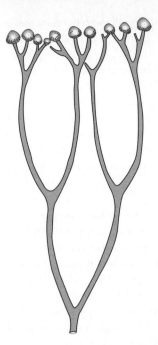

12. *Cooksonia*

spore-bearing organs, the presence of thickened walls of the vascular conducting tissues, and stomata on the outer surfaces of the stems. All these discoveries were won against considerable odds: much of the material is incomplete, and it all has to be processed through several elaborate protocols.

While stands of *Cooksonia* in the Silurian perhaps reached a height of 6 centimetres at most, little more than roughly cut grass, vascular plants became rather larger in the Early Devonian, 400 million years ago. These Early Devonian terrestrial settings are best known from an extraordinary fossil locality in Scotland, the famous Rhynie Chert. The locality is remote, but when fossils were first found there in 1914, they soon attracted intense attention. Not only were these some of the oldest plants yet found, but they were also diverse, and exquisitely well preserved. Further, here and there among the stems and stalks were small arthropods and other animals.

The Rhynie fossils have been preserved by flash silicification by hot springs. Recent work by Nigel Trewin and Clive Rice from the University of Aberdeen has shown that much of Scotland was an active volcanic zone at the time. Rhynie in the Early Devonian was like Yellowstone National Park today, with hot geysers erupting and immersing vegetation in silica-rich waters at a temperature of 35 °C – an ecosystem frozen (or rather boiled) in time. The Rhynie Chert is an unusual, hard, flinty rock, speckled black and white. The fossils cannot readily be seen on the surface, and they have to be studied in cross-sections cut through the rock, polished to an exquisite thin lamina on a microscope slide, and examined at high magnification.

The Rhynie fossils include remains of seven vascular land plants, as well as algae, fungi, one species of lichen, and bacteria, as well as at least six groups of terrestrial and freshwater arthropods. What is amazing is the quality of preservation: every cell and fine detail can be seen, as if frozen in an instant and preserved forever.

The Rhynie ecosystem was no towering forest. If you went for a stroll in Scotland in the Early Devonian, the green rim of plants probably did not extend far from the sides of ponds and rivers, and the tallest plants would barely have brushed your knees (Fig. 13). To see anything, you would have had to go down on your hands and knees, and peer at the stems through a magnifying glass. Most of the taller plants had smooth stems, and branched simply in two, with knob-like spore capsules at the tops of their stems – just larger examples of plants like *Cooksonia*. *Asteroxylon* had small scale-like leaves growing up from the stem. Microscopic cross-sections of these plants show they had simple vascular canals, stomata, and terrestrial spores. Between the plants crept spider-like trigonotarbids and insect-like arthropods, and some of

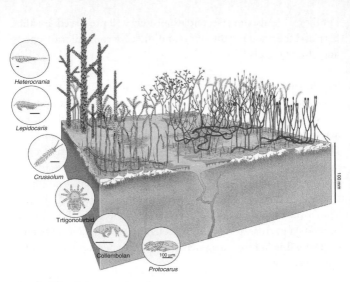

Heterocrania

Lepidocaris

Crussolum

Trtigonotarbid

Collembolan

100 μm

Protocarus

13. The Rhynie ecosystem

these are even found within cavities in the plant stems. There were crustaceans in the warm pools.

Through the remainder of the Devonian, mosses and other bryophytes lived in damp places, and did not seem to change much. But the vascular plants were evolving fast. They occupied spots further and further away from the waterside. The Rhynie plants all had their toes in the water, being connected to horizontal rhizome systems at the base, and the rhizomes were probably generally beneath water or damp mud. As the Devonian progressed, more and more vascular plants evolved their own root systems, and so were less dependent on standing water. The roots sought moisture at depth, and water permeated the plant through transpiration. These changes enabled the later Devonian plants to become larger than mere reeds, and some were positively tree-like by the end of the period, just before the time of great rainforests, the Carboniferous (see p. 87).

Scurrying through the undergrowth

Palaeontologists have hunted actively for evidence of animal life on land in association with the Ordovician and other very early soils and spore accumulations. So far, no luck. But there is some intriguing evidence that quite large animals actually moved on land at this time. In 2002, Robert MacNaughton from the Geological Survey of Canada, and colleagues, reported some large tracks produced by arthropods in desert-like sandstones dated as latest Cambrian or earliest Ordovician. These tracks are up to 29 centimetres wide, and they show symmetrical V-shaped markings probably formed by the back end of the animal, perhaps some sort of mollusc or worm that ploughed through the surface sands. An animal of such a size on land in the latest Cambrian is a real surprise!

The oldest body fossils of land animals come from the Late Silurian of Scotland and the Welsh Borders of England. In a series of studies, Paul Selden of Manchester University, and colleagues, have identified numerous land-living arthropod species from cuticle fragments. The fossils are microscopic, and they occur in black, organic-rich mudstones, and are not immediately obvious. The researchers break up the sediment, treat it with acid, often hydrofluoric acid, in order to break down all the sand grains and non-organic detritus. They can then pick through the plant and arthropod cuticles under the microscope. These studies were a huge surprise, because Selden and colleagues were able to extract diagnostic elements that matched bits of modern arthropods – pieces of legs, head shields, body segments, and other detritus.

The earliest land-living arthropods include millipedes and trigonotarbids. Millipedes are familiar enough today, but trigonotarbids are less well known. These are extinct spider-like arthropods, with eight legs (like modern spiders), and some at least could probably spin silk from spinnerets located at the back of the abdomen (also like modern spiders). Trigonotarbids did not have the narrow constriction between the abdomen and head portion, as seen in spiders, but a more straight-sided beetle-like body. Trigonotarbids were hunters, and they apparently lurked in and around the earliest plants awaiting their prey.

As mentioned above, the Rhynie Chert, somewhat younger, has also proved to be a rich source of early land animals (Fig. 13). The cherts have produced millipedes and trigonotarbids, as seen in the Late Silurian localities, but also shrimps and collembolans. Collembolans, more commonly called springtails, are strange little creatures, close to being insects, but not quite, that have a forked structure under their belly that is bent forwards. If threatened, the collembolan can release the fork and it propels the animal through the air for up to 80 times its own length in a split second – a good defence against a doubtless startled trigonotarbid or spider.

By the Middle Devonian, some additional modern arthropod groups had appeared. The Gilboa locality in New York State has produced fossils of millipedes, trigonotarbids, and mites, as well as the first insects and spiders. Scorpions are known from freshwater deposits before this time, but the first terrestrial forms occur in the Middle Devonian. Other arthropods from these sites include the long-legged predatory chilopods and the detritus-eating diplopods.

The Late Silurian and Devonian terrestrial ecosystems were different from today. The arthropods were mainly detritus-eaters and carnivores, with very few, or no, herbivores. Herbivory requires specialized gut microbes that can digest cellulose and lignin from plants, and this capability does not seem to have existed in the Silurian and Devonian. This is such a difference from modern arthropod communities, which are dominated by insects, and in which there are numerous forms that eat leaves and wood – think of all the larvae that eat plants in the garden and termites that can eat a wooden house in weeks!

So far we have seen life close up among the undergrowth. The arthropods of Silurian and Devonian times were all small, only a few millimetres in length. They were accompanied by worms and snails, although fossil records of these groups are sparse. Sooner or later, however, something larger was bound to make its ponderous way onto land and start eating all these nutritious little worms and arthropods.

The first tetrapods

Vertebrates may be divided loosely into fishes and tetrapods. Fishes have fins and swim in the water, and tetrapods (literally 'four feet') have legs and walk on land. Modern tetrapods are the amphibians, reptiles, birds, and mammals, but the first tetrapods were quite different from anything now living. The transition from fish to tetrapod happened in the Devonian, a time when coral reefs flourished in tropical seas, and fishes, many of them armoured, swam in shallow seas. Indeed, some of the Devonian armoured fishes were quite astonishing – *Dunkleosteus* was a placoderm ('platy skin') that reached a length of 10 metres, and could have engulfed anything in its vast jaws. Other fishes, though, had lungs and muscular fins that were used for dragging their bodies along lake beds.

A remarkable discovery in 2006 shed new light on the transition from fish to tetrapod. Three skeletons were retrieved from Devonian rocks of Arctic Canada that looked like hefty fishes (gills, scales, streamlined skull), but had some tetrapod characters (powerful limbs with rotating wrist and ankle joints, mobile neck, weight-supporting ribs). This creature, named *Tiktaalik*, was clearly capable of hauling itself onto land and breathing air as it sought another pond.

The next step is seen in fossils from the Upper Devonian of Greenland, some 370 million years old, that came to light in 1929. These were later named *Acanthostega* and *Ichthyostega* (Fig. 14).

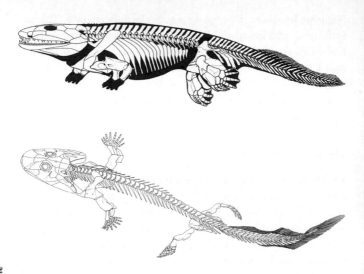

40mm

14. *Ichthyostega* and *Acanthostega* reconstructions

They both measure 0.5–1.2 metres long, and they were carnivorous, presumably feeding on fishes. *Acanthostega* and *Ichthyostega* retain a fishy body outline with a streamlined head and a tail fin. The skull is very like that of their fishy ancestors, being smoothly streamlined, and carrying lateral line canals. These structures are found in many Devonian fishes and in life presumably carried nerves and sensory organs that detected movements underwater. Modern fishes have such structures.

The main differences from fishes are seen in the limbs and the limb girdles. In fishes, the shoulder girdle attaches to the back of the skull, behind the gills. This strengthens the front of the body and provides a firm anchor for the *pectoral fins*, the front pair. In an animal that walks on land, having the shoulder girdle fused to the skull might cause problems: as the animal walked it would jolt its head all the time, and this would disturb its hearing and other senses (and might make it bring up its lunch). The *pelvic girdle*, the hip region, is fused to the backbone on each side, and this provides an even firmer anchor for the hind limbs.

The limbs of *Acanthostega* and *Ichthyostega* are the key to everything. As expected, they are just like our arms and legs – a single upper element, the humerus in the arm and the femur in the leg, a pair of lower elements in the forearm (radius and ulna) and shin (tibia and fibula), various wrist and ankle bones, and the fingers. But how many fingers?

Mike Coates and Jenny Clack, at the University of Cambridge, had a surprise when they prepared the hand region of one of their specimens of *Acanthostega* in the 1990s: they found that it had eight fingers. They then investigated the hindlimb, and found that it had eight toes. The classic specimens of *Ichthyostega* actually showed seven fingers and toes, and *Tulerpeton*, a relative from Russia, has six.

So this means that five digits are not fundamental to tetrapods. Humans have made much of the pentadactyl (five-digit) hand and foot – indeed it's the basis of the decimal system. Had we retained eight, seven, or six digits, perhaps our mathematics would be rather different. And, as for pianos and clarinets, who knows? Coates and Clack were very clear about what this meant: there is nothing fundamental about five digits, and indeed modern work in developmental biology shows that this is true.

In development, a very early embryo has no limbs. Then small, featureless limb buds appear, the rudiments of the arms and legs. As the embryo grows larger, the limb buds extend and differentiate. The single bone of the upper arm and thigh appears first, then the double elements of shin and forearm, and finally the wrist and ankle elements. The fingers and toes pop out in sequence, but the numbers are not predetermined. The effects of different developmental genes, interacting with the tissues of the developing embryo, fix the number. So, early tetrapods experimented with many different numbers of digits, and five became more or less the norm by the end of the Devonian. But of course many tetrapods today have fewer (frogs have four, rhinos have three, cattle have two, and horses have one). No tetrapod with limbs has entirely dispensed with toes and fingers.

The Late Devonian tetrapods were still aquatic, as shown by the tail fin, lateral line system, and internal gills. The vertebral column was flexible, as in a fish, and *Ichthyostega* and *Acanthostega* could have swum by powerful sweeps of their tails. The limbs are still oriented more for swimming than walking, and the hands and feet, with seven or eight digits, are broad paddles. If this is true, and Mike Coates and Jenny Clack believe the case is very clear, we have to look at the origin of terrestrial habits in tetrapods rather differently from the standard model (that is, fishes stepped onto land and stayed there).

Mike Coates has argued that *Acanthostega* lived most of the time in stagnant, vegetation-choked backwaters, emerging in damp conditions, but staying underwater in the dry season and gulping air at the surface. It walked largely underwater, stepping over vegetation, and kicking itself along the bottom. So, it could be that these Devonian tetrapods had indeed stepped out of the water for a while, and then reverted to a somewhat more aquatic existence, or they might never have made the move completely onto land. This certainly happened in the Carboniferous, when tetrapods diversified and some became quite dedicated land animals.

Being a land animal . . .

Arthropods and tetrapods faced different challenges when they moved out of the water. For the tetrapods, the main problem was weight and structural support, whereas for the smaller arthropods, this was probably not much of an issue. In addition, both groups had to evolve new modes of locomotion as well as new ways of feeding, of sensing prey and predators, of water balance, and of reproduction. Air breathing, as mentioned before, was a relatively minor problem.

With all these problems, it might seem a wonder that animals bothered to depart the safety of the Silurian and Devonian waters and venture onto land. Alfred Sherwood Romer, the great doyen of mid-twentieth-century vertebrate palaeontology, argued that vertebrates moved onto land in order to get back into the water. This isn't such a paradox: he argued that the Devonian was a time of seasonal droughts, and the freshwater fishes probably found themselves often in stagnant and dwindling pools. Then, those that were able to gasp a few breaths in air, and haul themselves with effort along a watercourse to the next pool, would survive. The fishes that could not cope out of water then died.

Romer's idea has been criticized because there is actually only limited evidence of droughts in the Devonian (Romer was no sedimentologist), and the 'getting back to water' model would not explain why tetrapods continued to hone their land adaptations. It's more likely that the move to land by both arthropods and vertebrates was simply to exploit new opportunities. There were plants for the arthropods to hide in and whose detritus formed the base of a food chain. Once the arthropods were there, the tetrapods doubtless followed them and gorged themselves on juicy millipedes and trigonotarbids.

Structural support was the key issue for the first tetrapods, as mentioned. A fish is buoyed up by the water and its body weight is pretty much zero. On land, however, the body becomes heavy, and the belly has to be held clear of the ground if the animal is to move forward without wearing its ventral surface away. In addition, the internal organs weigh down within the rib cage, and there's a risk of suffocation or damage. The whole skeleton has to become modified to counteract gravity, to hold the guts in, and to allow the animal to hoist itself up and propel itself forwards.

Tetrapods move in a very different way from fishes in water. Instead of a smooth gliding motion driven by sideways beating of the body, the limbs have to operate in a jerky fashion, producing steps. The fishes most closely related to tetrapods are the sarcopterygians, or lobe-fins. Living sarcopterygians include the lungfishes of southern continents, and the famous 'living fossil' *Latimeria*, the coelacanth. Devonian sarcopterygians were a diverse group, and they all had muscular fins containing bones, and they could all have 'walked' on the bed of a pond by stilting themselves forwards. It then took only a moderate amount of evolution for the muscular fins of a lobefin to become an arm or a leg.

The earliest tetrapods also had to modify the ways in which they fed and breathed. The skulls of the ancestral sarcopterygian fishes were highly mobile, but this ability was largely lost in the early tetrapods. The jaw movements of tetrapods are much simpler than those of most fishes, and they could just snap at prey, and not chew it. Air-breathing requires lungs, but the sarcopterygians already had lungs. Lungfishes today can breathe with their lungs in the air, but they can also absorb oxygen from the water through their gills, and through other tissues in the mouth. Doubtless, the first tetrapods could respire in several ways also.

Sensory systems had to change too in the first tetrapods. The lateral line system could only be used in the water. Eyesight was as important on land as in shallow ponds, and the sense of smell may have improved, but there is no evidence of that in the fossils. Early tetrapods had a poor sense of hearing in air, as did their ancestors – we know this because their main hearing bone, the stapes, which connects the eardrum to the brain, was a rather massive rod of bone, and was surely not capable of discriminating subtle differences in sounds.

. . . but only a partial solution

The first arthropods seem to have become more or less adapted to life on land at their first effort. They had waterproof cuticles, and some at least presumably laid their eggs on land. But this was not the case for the first tetrapods: they cracked most of the business of living on land, but left a couple of problems unresolved.

The first unsolved problem was the maintenance of water balance. In the air, water can evaporate through the moist skin of the body, the lining of the mouth and nostrils, and the early tetrapods risked desiccation. The earliest tetrapods probably remained close to fresh water, which they could drink in order to avoid this problem. Later, reptiles evolved waterproof scales and skins so they could escape entirely from the water, and still avoid desiccation in even the driest of conditions. The first tetrapods were certainly not capable of that.

Reproduction was the final hurdle to be crossed, and the first tetrapods, and amphibians today, made no evolutionary headway with this at all. Modern frogs and salamanders lay their eggs in ponds, and the young hatch as tadpoles. Tadpoles are really little fishes that live entirely in the water, and it is only after metamorphosis that the adult amphibian enters into a more terrestrial life, but even then only in a rather uneasy fashion. We know that early tetrapods had the same double mode of reproduction because a number of fossil tadpoles have been found. Again, it took some time before the reptiles came on the scene, and finally solved the terrestrial reproduction problem by producing eggs with shells that could be laid on dry land.

The Carboniferous Period followed the Devonian, and this was the time of the great coal forests. Not only were these the source of coal, and so of the industrial revolution and the modern world, the Carboniferous was also a time of rapid change in terrestrial ecosystems, and the world began to take on something like its modern appearance for the first time – but that is true perhaps only when viewed from a distance. In closer focus we might be startled by 2-metre-long millipedes, dragonflies as large as seagulls, and great trees that looked more like ferns than anything we are familiar with.

Chapter 5
Forests and flight

In Carboniferous forests dragonflies grew as big as ravens. Trees and
other vegetation likewise attained outsized proportions...
Bill Bryson, *A Short History of Nearly Everything* (2003)

These are the lasting images of the Carboniferous – great forests of
strange fern-like trees, and huge insects flying between their
trunks. The other image of course is of the legacy of those lush
forests, the vast coal mines where tons of coal are stripped from
the surface or hewn from coal seams at depth. The Carboniferous
was a crucial time in the evolution of life on land. This was the
time when plants and animals really cemented their land-living
adaptations, and took on all habitats and all continents. The rapid
rise of insects, tetrapods, and plants marked the future structure
of terrestrial ecosystems.

Life in the sea was no less rich. Tropical reefs abounded, some of
them a kilometre or more in length, and composed of dozens of
species of corals. Brachiopods, molluscs, and echinoderms lived
among those reef organisms, and conical and coiled molluscs
swam above side by side with sharks and other fishes, some of
them like modern forms, others much more weird and wonderful.
Some Carboniferous sharks were long and thin, others were
deep-bodied, some had long pointed snouts, others had great coils

15. A Carboniferous riverbank

of teeth at the front of their mouths, and some even had great bony spines covered in teeth that extended like sunshades from their foreheads.

But it is life on land we will explore here. The key biological inventions were forests and flight (Fig. 15). Before the Carboniferous, plants were sparse, and focused around water bodies, and animals crept around on the ground. Land life exploded in the Carboniferous, forests clothed much of the landscape, and numerous insects buzzed and whizzed in the sky. Was it pure chance that these changes happened at this time, some 320 million years ago, or was there something special about the Carboniferous world?

The world of the Carboniferous

The Carboniferous Period, from 360 to 300 million years ago, was a time of continental fusion. In the Devonian there had been several continents, a major northern continent that consisted of most of North America and Europe, as well as several southern continents. In the Carboniferous, these northern and southern landmasses began to fuse together, and indeed, in a line roughly along the modern Mediterranean, there was a great collision as Africa drove north into North America and Europe, causing earthquakes and volcanoes, and raising a chain of mountains from the Appalachians, across Ireland and Germany, to Poland. Much of Europe and North America lay around the Carboniferous equator, and tropical conditions prevailed throughout these regions.

Climates through the Early and Mid Carboniferous were warm, but conditions changed towards the end of the period. A huge glaciation began to develop at the South Pole. Our modern world is unusual in that we have an ice cap at both North and South Poles, and it is important to realize that there was no ice cap at either pole for much of the Earth's history. The absence of ice caps, as in much of the Carboniferous, means that there was much less temperature differentiation from the equator to the poles than there is today.

But why do we have ice caps today if this is not the norm? The general assumption is that ice can grow at the poles only if there is land at the poles. This is the so-called *albedo effect*, that cold begets cold. An ice cap of some size is necessary to set things going. Sunlight is reflected from white surfaces (such as ice) and absorbed by dark surfaces. So an ice cap is somewhat self-sustaining by being white; albedo is the extent to which an object reflects light, hence the term. If the ice cap sits on land, at one of the poles, or in the high mountains, the ice tends to remain and not melt, even under modest sunlight. At the poles, it is never really warm because of the angle of the Earth to the Sun, so the winter ice can survive all through summer. Antarctica sits squarely over the South Pole today, and Greenland is close enough to the North Pole to have the appropriate effect.

In the Carboniferous, and much of the history of the Earth, the poles of course were cold, but there was no land there. If oceans lie at the poles, the winter sea ice disappears each summer, and an ice cap cannot develop. The circulation and mixing of sea water also helps break up ice caps – deep cold waters move in stately fashion from the poles towards the equator, and rise, thereby pushing currents of warm water, ever so slowly, back towards the poles.

In the Mid Carboniferous, the southern supercontinent of Gondwana (including what is now South America, Africa, Antarctica, and Australia) was moving south, and it approached the South Pole. As this happened, an ice cap began to build up, and this survived for 30 or 40 million years, into the subsequent Permian Period, when the ice finally disappeared as Gondwana drifted away from the pole. There is extensive evidence for this southern glaciation, both geological and palaeontological. For a time, there was no life around the South Pole – the coals disappeared – and the rocks show clear evidence of glaciation: glacial tills and rock scratching, as seen earlier in the Neoproterozoic (see pp. 47–9), as well as sands compressed and contorted by the weight of ice, and scattered *erratic* ('wandering') blocks, rocks that had been torn up by glaciers and dropped elsewhere randomly.

Early observations of the evidence for Carboniferous glaciation were key elements in proving the concept of continental drift. About 1900, geologists pooled their evidence for glaciation in Australia, South Africa, South America, and India. They noted that the glacial features all pointed back to a source of ice in South Africa. Glaciers had evidently radiated from there, eastwards across Australia, westwards across South America, and northwards across India. On the modern map, India of course lies in the northern hemisphere, so how on earth, these early geologists asked, could ice have flowed all the way from the South Pole and across the equator?

Coal

The Carboniferous is famous for its coal. Indeed the name Carboniferous comes from the French *carbonifère*, meaning 'coal-bearing'. Huge coal deposits occur in Mid and Upper Carboniferous rocks of Europe and North America. Coal is almost pure organic carbon, composed of the remains of tree trunks, leaves, branches, and other plant debris that have been buried and compressed. Sometimes these plant remains may still be seen in the coal; in high-grade coals such as anthracite, the plant remains have been compressed and heated at depth and very little trace of their structure remains.

It is a puzzle why there is so much Carboniferous coal, and not much from other times. There are indeed some coal deposits from younger rocks, and some of these are commercially exploited, but these deposits pale into insignificance beside those from the Carboniferous. The clue may be partly in the worldwide models for the accumulation of coal: because of the particular Earth movements at the time, and perhaps because of the southern polar ice cap, land was subsiding rapidly in many areas, and huge thicknesses of terrestrial sediments, including coal beds, accumulated. But why were lush forests so extensive then?

The clue might come from studies of ancient atmospheres. It might seem incredible that geologists can reconstruct ancient atmospheres, or indeed that ancient atmospheres might have been different from today's. We have seen that the earliest Earth had an atmosphere that was devoid of oxygen (see p. 39), and that oxygen levels built up to near-modern levels by the end of the Precambrian. However, there is extensive evidence that levels of oxygen and carbon dioxide have varied considerably through the past 500 million years, and the Carboniferous was a time of extraordinarily high oxygen levels.

Geochemists focus in particular on isotopes of carbon and oxygen in the rocks. The secret is that the waters of seas and lakes contain similar proportions of gases to the atmosphere of the time, and these proportions may be locked into the skeletons of shellfish and planktonic organisms, as well as into limestones deposited on the seabed or certain kinds of ancient soils. It's essential of course to choose samples that have not been altered by any subsequent geological activity, so that the chemical measurements actually reflect conditions all those millions of years ago.

Many elements may exist as several *isotopes*, forms with different atomic weights. Isotopes are important in radiometric dating (see pp. 17, 20), and also in reconstructing ancient environments. Carbon, for example, exists generally as carbon-12 in living organisms, as carbon-13 in inorganic reservoirs, and even as the radioactive form carbon-14 in some settings. Nearly 99 per cent of carbon on Earth is in the form of carbon-12, and the other two isotopes make up much smaller quantities. Oxygen exists in three isotopic forms, oxygen-16, which is commonest, as well as oxygen-17 and oxygen-18. By measuring the ratios of these isotopes in ancient rock and fossil samples, geochemists can reconstruct ancient temperatures (from the ratio of oxygen-18 to oxygen-16), and detect perturbations to the carbon cycle (for example major extinctions, volcanic eruptions, releases of carbon from deep stores) from the ratio of carbon-13 to carbon-12.

All the measurements indicate that the Carboniferous atmosphere carried something like 35 per cent of oxygen, compared to 21 per cent today: this was the highest level that oxygen ever reached. The reasons are debated. It could simply be that the huge rise in plant diversity and abundance meant that global levels of photosynthesis increased, and so vast amounts of additional oxygen were pumped into the atmosphere. The problem with this idea is that oxygen levels fell to modern levels soon after the

Carboniferous, and yet plant richness remained as high as ever. Another reason might be that large quantities of wood were buried during this period because nothing could eat the lignin: today there are specialist bacteria that can reduce lignin rapidly, and wood-eating animals such as termites and beavers have such bacteria in their guts. If huge quantities of wood were buried, as they were, carbon was removed from the atmosphere through photosynthesis of carbon dioxide, and locked up in the buried, but not decomposed, plant cells, so releasing more gaseous oxygen and building up overall atmospheric oxygen levels.

Many palaeontologists have speculated that the extraordinarily high levels of oxygen in the Carboniferous might have permitted huge insects to evolve. Indeed, there were dragonflies like birds, cockroaches as large as your hand, 2-metre-long millipedes, and so on. Gravity was just the same in the Carboniferous, so why would insects be larger? There may be two reasons. Some physiologists have suggested that the oxygen-rich Carboniferous atmosphere would have been denser than today's, and so would have provided more lift and thus made it easier for them to fly. This sounds a bit weak, and in fact it probably wasn't the whole story.

The key may be in respiration. Insects 'breathe' by diffusion, and do not pump air, as we do, in and out of lungs. Insects have pores in their cuticle that extend deep into the body. Oxygen diffuses passively into their tissues through branching tubes, and this limits the size of an insect, and indeed the size of most arthropods. Because oxygen diffuses into the body, the cross-section of an insect is limited, and insects generally cannot be much larger than the largest dragonfly today, with a 15-centimetre wingspan, and a body as thin as a pencil. *Meganeura*, the Carboniferous dragonfly, had a 75-centimetre wingspan, and a body more like a frankfurter than a pencil. An atmosphere rich in oxygen may have been sufficient to allow arthropods to achieve gigantic size even with their passive diffusive system of respiration.

The great coal forests

Damp forests of vast trees and lush undergrowth became widespread in the Middle and Late Carboniferous. The plants included giant clubmosses, horsetails up to 15 metres tall, ferns, and seed-ferns (Fig. 15). There were no flowering plants – these came much later, in the Cretaceous (see pp. 15, 143) – and conifers were rare. These Carboniferous plants though had some of the first proper leaves – necessary to promote photosynthesis at a time when carbon dioxide levels were low.

Clubmosses were generally low shrubs, but some of them became huge in the Carboniferous. The best known is *Lepidodendron*, a clubmoss that reached 35 metres or more in height. Fossils of *Lepidodendron* have been recognized for 200 years because they are commonly found in association with commercial coalfields in North America and Europe. At first, the separate parts – roots, trunk, bark, branches, leaves, cones, and spores – were given different names, but over the years they have been assembled to produce a clear picture of the whole plant. The massive roots of *Lepidodendron* specimens may be seen *in situ* in a famous Victorian museum in Glasgow. When the specimens were found in an old quarry in 1887, they were carefully excavated, and covered over by a marvellous greenhouse constructed from cast iron and glass, and this can still be seen in the Victoria Park.

Horsetails are familiar to gardeners today as small pernicious weeds. Their upright green shoots, with a characteristic jointed structure, are linked by underground rhizome systems. Horsetails today may be rather obscure, and generally small, but this was a significant group in the Carboniferous that grew in incredibly dense, bamboo-like thickets. One form, *Calamites* reached nearly 20 metres in height, but shows the jointed stems and whorls of leaves at the nodes that are typical of modern smaller horsetails. The trunk of *Calamites* generally arose from a massive underground rhizome. The leaves formed radiating bunches at nodes along the side branch, and there were usually two types of cones.

The clubmosses and horsetails occupied the low-lying floodplains. Seed-ferns, conifers, and ferns were adapted to drier conditions, and they occupied elevated locations such as levees, the banks of sand thrown up along the sides of rivers. Ferns today are generally low-growing herbaceous plants, common in many environments. Some of the Carboniferous ferns were tree-like, with their fronds borne on a vertical trunk, while others were smaller, like typical modern ferns.

Seed-ferns were a diverse group of shrubs and trees. Conifers today include pine, spruce, and monkey puzzle. The Carboniferous cone-bearers, as today, were adapted to dry conditions. The earliest conifers had cones and long needle-like leaves that were adapted to save water. More familiar conifers only evolved rather later, in the times of the dinosaurs.

Some of the best evidence about Carboniferous plants comes, ironically perhaps, from burnt charcoal that was buried after wildfires. In the tropics today there are often vast wildfires that burn up hundreds of acres of forest. Fires may be started by a carelessly thrown cigarette or a bottle that focuses the rays of the sun, but usually the causes are natural; fallen branches and leaves may just be so dry that a chance lightning strike may spark off a huge conflagration that burns for days or weeks. The charcoal that is left after the burning still preserves all the fine cellular detail of the ancient wood, and ancient examples can reveal a huge amount of information.

Wildfires are not always destructive; indeed, many plants rely on occasional fires to clear old timber and to allow new shoots to grow. And the ash from the fire provides phosphorus and other nutrients. Wildfires were common in the past, and particularly in the tropical belt during the Carboniferous, and the high atmospheric oxygen levels surely stimulated more burning than happens today. Investigations show that fires were commonest in the higher areas, away from the banks of rivers, where plant debris could become very dry. Some of the Carboniferous wildfires might have been set off by nearby volcanic eruptions, or they might have been set off regularly during particularly hot or dry seasons. Wildfires were probably a regular part of the growth and regrowth of forests, and also destabilized hill slopes, triggering occasional landslides.

The Carboniferous wide-mouths

The new forest habitats of the Carboniferous opened up great possibilities for the early tetrapods, and they diversified extensively. As we have seen (p. 79), Late Devonian tetrapods are quite rare, and they still remained largely in the water. The groups diversified into some forty families in the Carboniferous, some of which continued to exploit freshwater fishes by becoming secondarily aquatic, while others became adapted to feed on the insects and millipedes.

There used to be a major gap in our knowledge of tetrapod evolution in the Early Carboniferous, but new work on localities in Scotland has revealed some extraordinary animals from this time. One of the strangest is *Crassigyrinus*. It has a large skull with heavily sculptured bones and massive jaws, which show it was clearly a fish-eater. In fact, *Crassigyrinus* was really a massive head driven by a bulky body and minute limbs, and it could barely have struggled about on land.

The whatcheeriids, known from Scotland and the United States, were metre-long animals, also with massive heads, but perhaps mixing their diet of fish with occasional tetrapods too. The baphetids, known from Europe, had very low skulls. Their broad curved jaws were lined with small teeth, and they may have hunted rather small fishes. This head shape, roughly semicircular, with broad grinning jaws, and a skull that is about the same depth as the lower jaws, was typical also of a major group of Carboniferous tetrapods, the temnospondyls. There were many different lines of temnospondyls, and the group diversified and was important through the Permian and Triassic, and survived right into the Cretaceous, some 200 million years later. Most temnospondyls were up to a metre in length, and they were generally fish-eaters, but some were much larger; a few much smaller.

The most unusual Carboniferous tetrapods were the lepospondyls. There were three lepospondyl groups, the small insect-eating and quite terrestrially adapted microsaurs, the newt-like nectrideans, some with extraordinary broad heads shaped like boomerangs, and the enigmatic limbless aïstopods. The microsaurs dashed about on dry land, the nectrideans were strongly adapted to life in the water and may have snapped at insects above the water, while the aïstopods perhaps hunted slugs and worms in the damp leaf litter of the forest floor.

All the tetrapod groups mentioned so far were on the amphibian side of the fence, and it is probable that modern amphibians evolved from this group. The oldest frogs are Triassic in age, and probably evolved from among the temnospondyls. Salamanders might have arisen at the same time, but their oldest representatives are known first from the Jurassic.

The other major tetrapod branch were the reptiliomorphs ('reptile-like'). Basal reptiliomorphs included the Carboniferous anthracosaurs, a group of medium-sized rather aquatic animals that chased fishes in the rivers and ponds. But, midway through the Carboniferous, the reptiliomorphs spawned something surprising: the first reptile.

Reptiles and eggs

The great Scottish geologist Sir Charles Lyell (1797–1875) had the surprise of his life when he visited the wind-lashed shores of Nova Scotia in 1852. He had been there before, in 1842, a hazardous and brave journey for a geologist more used to the quarries and coasts of Europe and the debating rooms of the Geological Society in London. Lyell had written his epochal *Principles of geology* in the early 1830s, and these had set the new science of geology on a firm road for its future development. Lyell was committed to understanding how the Earth worked, and he undertook travels to remote continents to augment his understanding and to provide materials for his hugely popular textbooks.

In 1852 Lyell was exploring the cliffs at a locality called Joggins, on the north shore of Nova Scotia, in company with the Canadian geologist William Dawson (1820–99) who had earlier found some remarkable specimens of tetrapods preserved in an ancient tree trunk. Lyell was amazed by what he saw. A tree trunk stood there in the cliff, upright, in the position of growth. Modern erosion by the sea had worn away the rock and the two geologists could see inside the ancient tree stump. There, within the sand inside the trunk, were the tiny bones of a tetrapod. Lyell later reported that the Joggins locality was 'the finest example in the world'.

What Dawson and Lyell had found was a specimen of the oldest fossil reptile, later named *Hylonomus*. This animal was about 30 centimetres in length, roughly lizard-shaped, with a long tail and long limbs. Its small sharp teeth show that this was an insect-eater. The specimens of *Hylonomus* are superbly preserved because they have sat undisturbed within the ancient tree stumps; after 1852, many more examples of this stump style of fossil preservation have been found at the Joggins Cliff. It was only in the 1960s, over 100 years after its first discovery, that palaeontologists realized that *Hylonomus* was not a microsaur, but actually a reptile. Its high skull sets it apart from most of the Carboniferous tetrapods, which had low skulls. But more importantly, *Hylonomus* has a major ankle bone called the *astragalus*, a bone not seen in amphibians, but typical of reptiles, birds, and mammals.

How did these animals die? It seems that the trees were felled by a sudden flood in the Mid Carboniferous, and the trunks and branches were swept away. The tree stumps were presumably firmly enough rooted that they did not budge. The flood waters washed sand and mud around the tree trunks, partly burying them. As the core of the trunks rotted away, the woody tissues may have been attacked by insects, and the reptiles may have followed them in, looking for a snack. Why in the end the reptiles became trapped is not certain. Surely they could have climbed out of the hollow trunk? Or maybe a further flood swamped them before they could escape.

Why is *Hylonomus* so significant in evolution? The point is that it was the first tetrapod to lay eggs, and so to escape from the hold of the water. As we saw (p. 83), the first tetrapods, the amphibians, still relied on their proximity to the water to avoid desiccation and to lay their spawn. All other tetrapods, the reptiles, birds, and mammals, have made the break away from the water, and *Hylonomus* was the first. And yet there are no fossil eggs known from a Carboniferous reptile, so how do we know that *Hylonomus* and its descendants laid eggs?

The answer comes from phylogeny and the idea of homology (see p. 12). Modern reptiles, birds, and mammals lay very similar eggs, called *amniotic eggs*. The amniotic eggshell is usually hard and made from calcite, but some lizards and snakes have leathery eggshells. The shell retains water, preventing evaporation, but allows the passage of gases, oxygen in and carbon dioxide out. The developing embryo is protected from the outside world, and there is no need to lay the eggs in water, nor is there a larval stage in development. Inside the eggshell is a set of membranes that enclose the embryo, that collect waste, and that line the eggshell. The developing animal is sustained by the highly proteinaceous yolk.

Reptiles, birds, and mammals are called collectively the amniotes, or the amniota, because they all share the same ultimate ancestor, an animal close to *Hylonomus*. Birds evolved from dinosaurs in the Jurassic (see p. 138) and mammals evolved from another reptilian group in the Triassic (see p. 130). But do mammals lay eggs, as I have just claimed? Well, the most primitive living mammals, the platypus and echidna from Australia, do in fact lay eggs with a hard calcareous shell. The details of the anatomy of reptilian, avian, and mammalian eggs are all the same, and so these all evolved from a single ancestor. Track back down the evolutionary tree, and you arrive at *Hylonomus*, so *Hylonomus* must have laid the same kind of egg.

The Carboniferous then seems to have been some kind of acme, or high point, in the evolution of life both in the sea, and especially on land. Nothing could ever equal the high oxygen levels, the vast acreages of lush tropical forests, and the giant insects. Indeed, only 50 million years later, at the end of the subsequent Permian Period, the most devastating mass extinction of all time would kill off nearly all of life.

Chapter 6
The biggest mass extinction

> [Mass extinctions may operate by] the Field of Bullets scenario – all
> individuals exist in a field of flying bullets, and death or survival is
> only a matter of chance. The image is awful, but it does the job.
>
> David Raup, *Extinction: bad genes or bad luck* (1991)

Extinction isn't all bad. In fact, no species can last more than a few
million years at most, before it is supplanted by other species, or
evolves into something else. So extinction is happening all the
time – this normal kind of extinction is called *background
extinction*. There have been episodes in the history of the Earth,
however, when more extinction than is normal happens in a short
span of time. These times are called *extinction events*, and they
can include examples where all life on a particular island has been
wiped out by a local catastrophe, or where major climate change,
or hunting, kills off certain kinds of organisms, such as the
widespread extinction of large mammals at the end of the ice ages
11,000 years ago.

Most fascinating, and troublesome, are the big extinction events,
the so-called *mass extinctions*. These are times when much, or
most, of life disappeared at once. There are generally reckoned to
have been at least five mass extinctions, as follows (approximate
ages in millions of years): Late Ordovician (440), Late Devonian

(370), end-Permian (250), end-Triassic (200), and end-Cretaceous (65). These are known as the 'big five', and they are set apart from all other extinction events by three factors: more species died out during each event than at any other time, the victims were of diverse ecology and worldwide distribution, and there appears to have been a single major global crisis that triggered the event. Recent work by Dick Bambach of Harvard University, and colleagues, suggests that we perhaps ought to talk about just a 'big three' – the Late Ordovician, end-Permian, and end-Cretaceous – because the other two seem to have been more prolonged.

Of these five (or three) mass extinctions, we will look at the end-Permian event in some detail in this chapter, and the end-Cretaceous mass extinction later (see pp. 144–5). Briefly, the other three events were significant in their time. The end-Ordovician event seems to be associated with a short, sharp, ice age, and it saw the end of many classes of trilobites, brachiopods, corals, and others. The Late Devonian event seems to have lasted for several million years, and there were further losses among brachiopods, ammonoids (coiled swimming molluscs), corals, and the heavily armoured fishes of shallow waters. Finally, the end-Triassic event again seems to have lasted over a few million years, and hit the brachiopods and ammonoids particularly hard, as well as many land-living reptiles.

The end-Permian mass extinction is especially significant because it was by far the largest mass extinction of all time. It is estimated that up to 96 per cent of species died out, and this is then the nearest that life has ever come to total annihilation. The search for the pattern and causes of this largest-of-all events has, however, been fraught with difficulty. It is almost impossible to imagine what a 96 per cent loss of species might have been like: that is, only 4 per cent of species – fewer than one in twenty – survived. It's important to understand what life was like before disaster struck.

Life in the sea: before and after

Life in the shallow seas of the latest Permian was rich and diverse (Fig. 16A). Organisms were living on and in the seabed, as well as swimming above it, and what the diver in the Late Permian would have seen was probably superficially like a modern coral reef. The reef would have consisted of hundreds of species. The framework of the reef was built from sponges, corals, and bryozoans, animals that secrete a stony skeleton in which they live. Living on the dead corals were various clinging molluscs and worms.

A whole variety of snail-like molluscs, starfish, and shrimps crept among the coral fronds. The brachiopods ('lamp shells') were the most important shellfish; most lived fixed to the seabed by a tough stalk, and they fed by filtering tiny food particles from the water. Molluscs were rarer then than now: clams and snails, relatives of the swimming ammonoids that were more mobile than the brachiopods, and grazed on algae on the coral skeletons, or hoovered up organic matter from the mud.

Associated with the reef were various kinds of echinoderms, some fixed, some mobile. Sea lilies (more properly, crinoids) were hugely successful in the Palaeozoic, forming parts of the great reefs, either growing on and around the corals and sponges, or forming huge crinoid forests on their own. Typical crinoids look like plants: a long flexible stalk fixed down by a root-like structure, and a blob at the top of the stalk with tentacle-like arms shimmering in the currents above. Crinoids feed on small organic particles in the water, which they capture on their sticky arms and then waft down a central groove on the upper surface of the arm into the mouth that is located in the centre of the body. The body is the blob on top of the stalk.

And swimming above were jet-propelled nautiloids and ammonoids (relatives of squid and octopus), swimming

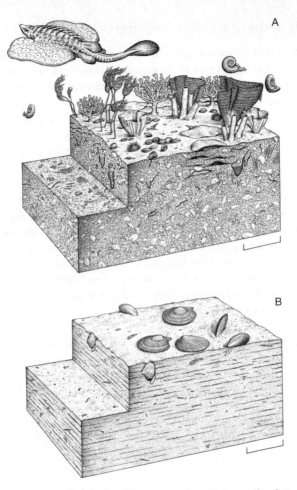

16. Life before (A) and after (B) the end-Permian mass extinction

arthropods, and fishes of various kinds, bony fishes and sharks.
There were also abundant microscopic organisms living as
plankton in the waters above the reef: radiolarians, with their
delicate net-like skeletons made from silica, and foraminifera, tiny
creatures with coiled and segmented calcareous shells. Just as

today, Late Permian reefs were real diversity hotspots, locations of unusual species richness.

All of this diversity was slaughtered by the end-Permian crisis. For example, fifty out of fifty-five brachiopod families died out (91 per cent loss). All but a tiny handful of this hugely diverse and abundant group bit the slime. Molluscs were less affected, but still suffered major losses, especially the ammonoids, which all but disappeared. The same devastation was meted out on the fixed reef-building organisms too: the Permian corals disappeared, and the bryozoans and crinoids were reduced to a handful of species.

Floaters and swimmers were particularly hard hit. Among the plankton, the radiolarians virtually disappeared. The foraminifera also suffered major losses. Among fishes, the diverse medium- and large-sized sharks of the Late Permian were reduced to a fauna of only small sharks in the Early Triassic. Whether it was only small forms that managed to survive, or whether the survivors, variable in size, then became dwarfed because of some evolutionary pressure is uncertain. Also, among the bony fishes, two out of eight families disappeared.

What about the post-extinction scene? In rocks only a few millimetres above the rich reef beds, there is a very different story (Figure 16B). The hundred or more pre-extinction species have been reduced to four or five. Most are particular kinds of bivalves, called paper pectens, that were attached by fine threads to irregularities in the black Early Triassic muds. Sharing the scene was a sudden horde of microgastropods, tiny snails, and *Lingula*, the old trouper. *Lingula*, we were taught as undergraduates, was a genus that had existed for 500 million years, from the Cambrian to the present day. It was an inarticulate brachiopod, meaning its hinge line lacked tooth-like locking mechanisms (of course, with the great originality of undergraduates, we used to quip that this was because brachiopods can't talk). *Lingula* is a simple brachiopod, consisting of two valves, both quite similar, and each shaped like a tear-drop, and it lived in waters of mixed salinity in coastal areas. Of course it is most unlikely that the one genus lasted so long: its anatomy is so simple, we probably just can't tell the difference between diverse genera and species. But this kind of brachiopod was certainly a survivor.

Oxic to anoxic

Marine conditions changed dramatically across the Permo-Triassic boundary, and this change might give crucial evidence about what caused the crisis. Geologists have now studied the rock sections across the Permo-Triassic boundary in many parts of the world, but one of the best successions is the Meishan locality in south China. Indeed, this rock sequence was selected in 2000 as the world type section for the base of the Triassic. This means that all geologists henceforth must use the Meishan section as their reference point.

Until 1990, there had not been much work on the Chinese Permo-Triassic sections, and such work was impossible during the Cultural Revolution. After 1990, it became easier for foreigners to work in China, and Paul Wignall from the University of Leeds and Tony Hallam from the University of Birmingham decided to go and have a look. Everyone said that working conditions in China would be impossible, and that the rocks weren't up to much anyway. Nevertheless, they persisted, and with a modest grant of a few thousand pounds from the Royal Society they went and they saw.

The Meishan section proved to be clear and straightforward. As sedimentologists, Wignall and Hallam were looking for evidence of ancient environments as recorded in the rocks. They noted many metres of thick and thin *bioclastic* limestones in the uppermost Permian; these are limestones made up largely from broken shells and other detritus of organisms. Such limestones attest to warm-water shallow seas, with some water currents that washed the shells and other animal remains around on the sea floor before they were finally incorporated into the rock. Near the top of the Permian, there is extensive burrowing in the limestones, indicating a fully oxygenated seabed. Then suddenly, everything changes. The thick, burrowed limestones disappear, and so too the abundant fossils.

The highest limestone is followed by 28 centimetres of clays and then another limestone. First, there is a pale-coloured ash and clay bed, then a dark organic-rich mudstone, and then a muddy limestone. In the Chinese system, these are numbered as beds 25, 26, and 27 respectively. Above bed 27 follows a long succession of thin limestones and black shales that contain only rare, small burrows. Here is a major succession of low-oxygen beds spanning one or two million years. Not only are the sediments black and devoid of fossils, but scatterings of pyrite crystals are found here and there.

Iron pyrites, or 'fools' gold', is iron disulphide (FeS_2). It forms when oxygen is absent from the decay of organic matter by anaerobic bacteria that convert sulphate to sulphide, with the loss of oxygen. Everyone has experienced the association of sulphide with *anoxia*, the absence of oxygen, when walking across a black muddy pool full of dead leaves. The foul rotten-eggs smell is sulphide. So, the earliest Triassic seabed at Meishan, and in most parts of the world, was anoxic.

In 1996, Paul Wignall and his student Richard Twitchett argued that the world's oceans passed through a global phase of anoxia at the beginning of the Triassic – superanoxia, as they called it. They have mapped all the sections people have studied so far, most of them falling around the coastlines of the supercontinent Pangaea, and found that everywhere became anoxic, except, for some unknown reason, a small patch in present-day Oman. Why the seas over Oman did not become anoxic is a mystery, but at least some species were able to survive in reduced numbers there.

The first marine faunas of the Triassic were not only much reduced in diversity, but they also became boring. Before the event, faunas showed regional *endemicity*: that is, they were different from place to place – just as is the case today, and as is normal for mature and undisturbed ecological communities. After the event, we find *cosmopolitanism* – all species were the same everywhere. The thin-shelled paper pecten *Claraia*, and the inarticulate brachiopod *Lingula*, spread around the world.

Physical changes

Geologists have long debated the causes of the end-Permian mass extinction, and ideas around 1990 focused on two phenomena: continental drift and vast eruptions in Siberia. The two great Carboniferous supercontinents, Gondwana in the south, and Laurasia in the north (see p. 90), had fused along a line crossing the Caribbean and Mediterranean areas of today. The single great supercontinent, Pangaea (literally 'all earth'), stretched from north to south, roughly equally balanced across the equator. Life on and around such a world is hard to comprehend. Species on land could then move widely from place to place without hindrance. Indeed, the Late Carboniferous polar ice cap (see p. 89) had by then disappeared, so global climates were equable from north to south.

Palaeontologists used to argue that the fusion of continents itself was a strong extinction mechanism. Surely, they argued, as continents joined together, regional faunas would mix, and diversity would be lost. This would have been especially true, it is assumed, for shallow seas. Many seas were closed off as continents fused, and all their biodiversity would have been squeezed out.

The problem with this idea is that it cannot generate a sudden mass extinction – continental movements are slow and stately, and any consequent extinction would last over tens of millions of years. Further, it's not clear actually how many species would be driven to extinction by such geographic processes. In more recent cases where continents have joined, such as the establishment of the Isthmus of Panama 3 million years ago, linking the continents of North and South America, there was an exchange of faunas, and some consequent extinction, but no real devastation.

A more likely cause of extinction was the eruption of the Siberian Traps. At the end of the Permian, giant volcanic eruptions occurred in Siberia, spewing out some 2 million km^3 of basalt lava, and covering 1.6 million km^3 of eastern Russia to a depth of 400–3,000 metres. It was first suggested in the 1980s that this massive volcanic activity might be linked to the PT mass extinction.

Early efforts at dating the Siberian Traps produced a huge array of dates, from 160 to 280 Mya, with a particular cluster between 260 and 230 Mya. More recent dating, using newer radiometric methods, yielded dates exactly on the boundary with a total range of 600,000 years, but further work has to be done to determine exactly how many major phases of eruption there were, and their precise dates. These can then be keyed to dated ash layers in sedimentary sequences as far away as southern China.

Life on land

Life on land was as rich as in the sea in the latest Permian. The best areas to study tetrapod evolution at this time are in the Karoo Basin of South Africa, and the South Urals region of Russia. In the Late Permian rock sequence in Russia, for example, skeletons of amphibians and reptiles are found abundantly through a succession of faunas that span the last ten million years of the Permian. The latest Permian fauna of Russia, the Vyatskian assemblage (Fig. 17), known from the North Dvina River and from the South Urals, was rich and diverse.

Vyatskian herbivores include the large pareiasaur *Scutosaurus*, a formidable hippo-sized animal covered in bony excrescences, and

17. Life on land in the Late Permian in what is now Russia

the large, smooth-skinned dicynodont *Dicynodon*, with its two expanded canine teeth and otherwise toothless jaws. Carnivores include four species of gorgonopsians, including *Inostrancevia*, a great sabre-toothed reptile that presumably preyed on *Scutosaurus* and *Dicynodon*, as well as two smaller carnivores, a therocephalian and a cynodont. In other localities, latest Permian reptiles include *Archosaurus*, a 1-metre-long slender fish-eating reptile, oldest member of the Archosauria, or 'ruling reptiles', the group that includes crocodilians and dinosaurs, and procolophonids, small triangular-skulled reptiles, related to pareiasaurs, but superficially looking somewhat like fat lizards. At the water's edge were three or four species of amphibians. This was a rich and complex ecosystem, with as many animals as in any modern terrestrial community.

The amphibians and reptiles that survived the crisis into the earliest Triassic in Russia are a poor assemblage, the so-called Lower Vetluga (Vokhmian) Community. The only reasonably sized herbivore was *Lystrosaurus*, and other tetrapods include one species of procolophonid, and some rare therocephalians and diapsids that fed on insects and smaller reptiles, as well as fish-eating, broad-headed amphibians.

The most prominent of these survivors on land was *Lystrosaurus*, a modest-sized dicynodont that is known first from the latest Permian in South Africa. The extraordinary thing about *Lystrosaurus* is not that it survived, but that it dominated the whole world for a short time. Species of *Lystrosaurus* have been described from South Africa, South America, Antarctica, India, China, Russia, and possibly from Australia – so this genus was cosmopolitan, just like *Claraia* and *Lingula* in the sea.

Lystrosaurus was not only cosmopolitan, but it was also hugely abundant. Over 2,000 skulls have been collected from South Africa, and collectors walk past any they see in search of the rarer forms. This one genus, in places, makes up more than 95 per cent of the fauna, a very weird imbalance and a far from natural community. Perhaps this excessive domination of earliest Triassic terrestrial faunas by one species indicates that something was awry, that normal ecological rules were not operating.

Was *Lystrosaurus* a super-organism? Perhaps it had hidden powers of some kind that enabled it to survive. Certainly, it seems likely that *Lystrosaurus* could burrow, and it may have had a rather broad range of plant foods in its diet. Both burrowing and a wide diet could be useful survival strategies in times of crisis. However, many other rather similar reptiles did not survive the end of the Permian, and it seems much more likely that *Lystrosaurus* was simply lucky rather than highly adapted. It survived where others didn't by chance (Raup's 'field of bullets' model), and then spread worldwide into other areas because nothing else survived there.

Timing of the event

Surprisingly for such a dramatic event as the end-Permian mass extinction, its duration has been hard to establish. Earlier estimates suggested that the extinction was really a long decline in species numbers through perhaps 10 million years of the late Permian, but more recent studies have shown that the event was rapid. The difficulty was partly because palaeontologists had not been precise enough in dating their fossils, and also that good radiometric dates for the Permo-Triassic boundary did not become available until the 1990s. The current most widely accepted date for the boundary, and the mass extinction, is 251 million years ago, although a strong minority prefers 253. The differences depend on subtleties of how the dating samples are studied, and this may be resolved shortly.

Equally important is to try to dissect the anatomy of what happened, and this requires studies of local sections. So far, the most comprehensive such study has been done on the Chinese

Meishan sections. Jin Yugan from Nanjing and colleagues presented a study in 2000 of fossils below and above the Permo-Triassic boundary at Meishan. They identified 333 species belonging to fifteen marine fossil groups – microscopic foraminifera, fusulinids, and radiolaria, rugose corals, bryozoans, brachiopods, bivalves, cephalopods, gastropods, trilobites, conodonts, fishes, and algae.

In all, 161 species became extinct below the boundary beds (Fig. 18), during the four million years before the end of the Permian. Extinction rates at individual horizons amounted to 33 per cent or less. Then, just below the Permo-Triassic boundary, at the contact of beds 24 and 25, most of the remaining species disappeared, giving a rate of loss of 94 per cent at that level (Fig. 18, level B). Then there is a stretch of rocks that shows species originating and disappearing before, between beds 28 and 29, something changes, and species seem to live longer and extinction rates drop off.

This interval, encompassing beds 26 to 28, is something special. There are volcanic ash beds at both bottom and top of the succession, giving dates of 251.4 and 250.7 million years respectively, a difference of 700,000 years. The radiometric daters tell us their dates are now rather precise, so we can perhaps believe that this is really a time span of half a million years or so. Note that this time interval matches the span of time involved in the eruption of the Siberian Traps, estimated at 600,000 years.

But what was going on in this interval? Perhaps, after the major extinction at the top of bed 25, we are seeing *disaster species* coming in, species that evolve quickly to suit the strained conditions of the time, but do not last long. Then, from bed 29 (Fig. 18, level C), species originate and do not go extinct rapidly, and it seems we have returned to something like normal conditions of life. This pattern of sudden extinction, and half a

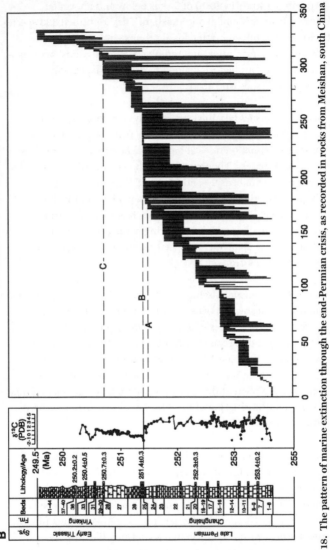

18. The pattern of marine extinction through the end-Permian crisis, as recorded in rocks from Meishan, south China

million years of nastiness, may be a further clue to what happened 251 million years ago. What about conditions on land?

Rivers and shock erosion

Since 1993, teams of researchers from the Universities of Bristol in the UK and Saratov in Russia have been investigating the Permo-Triassic boundary in Russia. In early work, we observed some remarkable sedimentary phenomena just at the boundary: it seems that the conditions of geomorphology and river patterns changed dramatically. Could this be a local phenomenon, or perhaps something more significant?

Valentin Tverdokhlebov, leader of the expeditions, had noted in the 1960s that the rate of sedimentation by rivers increased enormously in the earliest Triassic, and he attributed this to renewed uplift of the Ural Mountains. Vast alluvial fans spewed westwards from the west side of the Ural Mountains, each fan spreading for a length of 100 to 150 km over the low-lying Permian lakes and meandering rivers on the great plain. *Alluvial fans* are found today at major changes in slope along a river, especially where mountain streams, flowing at speed and carrying masses of boulders and coarse sediment, suddenly slow down as the river moves onto the flatter gradient of a plain.

Independently, Roger Smith, a sedimentologist working in South Africa, and his collaborator Peter Ward from the University of Washington in Seattle, had reached the same conclusion. The famous Permo-Triassic succession of the Karoo Basin showed a similar sedimentary switch from a low-energy flow regime with meandering streams in the Late Permian to a high-energy flow regime with braided streams and alluvial fans in the Early Triassic. Since then, this shift in fluvial style has been noted across the Permo-Triassic boundary in Australia, India, and Spain. Such a shift does not occur everywhere. Studies of soils of this age confirm that there was a soil erosion crisis, where soil and organic matter from the land was washed into the sea. If this was a worldwide phenomenon, then local-scale tectonism cannot be the cause – but what then?

Andy Newell, one of our team members, argued that the abrupt increase in channel size associated with a major influx of gravel around the Permo-Triassic boundary could be related to climate change. There was a well-documented switch worldwide from a semiarid/subhumid climate in the latest Permian toward one of greater aridity in the earliest Triassic, and this can increase sediment yield by reducing vegetation cover.

This model fits with other evidence that the normal green plants had been temporarily killed off, and replaced by an unusual horizon at the boundary that is dominated by strands produced either by fungi or algae. Below this horizon, the sediment samples contain spores of ferns, seed-ferns, horsetails, and other plants that grew at low, medium, and tree-like levels. Such plants soon return in higher units in the Early Triassic. But the fungal/algal boundary bed perhaps indicates a dramatic loss of normal vegetation. We know the devastating erosion that can follow the removal of plants today, such as in Bangladesh, where the rate of runoff and erosion has increased hugely after logging higher in the foothills of the Himalayas.

Soil stripping and massive sediment runoff are further clues to the nature of the crisis. The final evidence comes from isotopes.

Isotopes and climate change

Stable isotopes of carbon and oxygen are increasingly important tools in interpreting past conditions on the Earth, as we have seen (p. 92). At the Permo-Triassic boundary, there is a dramatic decrease in the value of the oxygen-18 ratio that corresponds to a global temperature rise of about 6 °C.

Global warming can cause anoxia by reducing ocean circulation and so reducing the amount of dissolved oxygen in the oceans. Lack of oxygen worldwide would surely have killed much of the life of the oceans, as Paul Wignall and Richard Twitchett had suggested.

Carbon isotopes have also been hugely important in determining models for the end-Permian mass extinction. The key isotopes here are carbon-12, which is characteristic of plants and animals, and carbon-13, which is found in inorganic settings. The Permo-Triassic boundary is characterized by a negative shift in the carbon-13 ratio: that is, an increase in the relative proportion of the carbon-12 isotope. On the face of it, this should imply a massive increase in the rate of burial of organic matter – perhaps the dead plants and animals killed by the extinction.

But this apparently is not enough. The amount of negative shift in the carbon-13 isotope ratio (4–6 parts per thousand) is too great to be explained solely by massive burial of dead organisms. An additional input of light carbon to the ocean-atmosphere system is required. The carbon dioxide emitted by volcanoes has low carbon values, but all the carbon dioxide produced by the eruption of the Siberian Traps would still not have been enough to cause the shift seen at the boundary.

The only viable source of sufficient light carbon might then be the methane trapped in gas hydrate deposits. Gas hydrates are great stores of organic carbon, formed largely from methane locked into ice and accumulated in deep oceans off the continental margins and in the permafrost of polar regions. The methane has formed from decaying organic material, dead plankton in the sea, and plant roots and leaves in the tundra. When the air over the tundra or the waters of the deep oceans warms up, the ice melts and the methane may be released rapidly. This is a once-only phenomenon, however. Once the gas hydrate methane has been released, it takes many thousands of years to recharge the reservoirs.

The runaway greenhouse

Is there a model for the end-Permian mass extinction that can produce such a devastating level of killing, and explain the timing over half a million years, the isotopic evidence, the anoxic seabed sediments, the stripping of vegetation from the land, and the light carbon isotopic shift?

Some researchers have suggested there was an extraterrestrial impact, as there was at the end of the Cretaceous (see pp. 144–5), but evidence is scant. More consistent with the evidence, but by no means proved, is an Earth-bound model that stems from the combination of the geological and palaeontological data already described, together with the fact that there were massive volcanic eruptions in Siberia at the same time.

Since 1990, attempts have been made to link the geological evidence for oceanic anoxia, global warming, and a catastrophic reduction in the diversity and abundance of life with the eruption of the Siberian eruptions to provide a coherent killing model. The sharp negative excursion in carbon isotope values implies a dramatic increase in the light isotope carbon-12, and geologists and atmospheric modellers have broadly accepted that there was a combination of sources, from buried organic matter, volcanic carbon dioxide, and methane from gas hydrates.

The assumption is that initial global warming at the Permo-Triassic boundary, triggered by the huge Siberian eruptions, melted frozen gas hydrate bodies, and massive volumes of methane rose to the surface of the oceans in huge bubbles. This vast input of methane into the atmosphere caused more warming, which could have melted further gas hydrate reservoirs. So the process continued in a positive feedback spiral that has been termed the 'runaway greenhouse' phenomenon. Some sort of threshold was probably reached, beyond which the natural systems that normally reduce carbon dioxide levels could not operate. The system spiralled out of control, leading to the biggest crash in the history of life.

Volcanic eruptions produce carbon dioxide, as well as other gases, and these, when mixed with water, turn into acids. So *acid rain* was an immediate consequence of the massive Siberian Trap eruptions, and this would have killed off land plants, whose debris was stripped and washed away, together with the soil on upland areas. Global warming, triggered by the excess carbon dioxide and methane being pumped into the atmosphere, led to stagnation of the oceans and seabed anoxia that lasted for some time.

There is evidence then that land plants were killed by acid rain, and that marine animals were killed by a lack of oxygen. The atmosphere was also short of oxygen, and this would have created major physiological stresses for many land animals too. It seems likely that the anoxic oceans became overloaded with hydrogen sulphide. There is limited evidence for this in the frequent occurrence of pyrite, a product of bacterial decay of organic matter in the presence of sulphates in the sea water. Sulphidic waters from the deep oceans could have risen higher and higher, replacing normal waters, and killing everything in their path, even releasing hydrogen sulphide into the atmosphere. So, if the land animals were gasping for air at lower-than-normal oxygen levels, the rotten-egg hydrogen sulphide gas could have finished them off.

Recovery

Aspects of the killing model are still highly speculative, and each researcher has his or her own hobby horse. But there is increasing hard evidence for several aspects of the ghastly scenario. After driving life down to 4–15 per cent of its previous diversity, how long did it take to recover?

Isotopic evidence shows that there were repeated carbon anomalies for the first 5 million years of the Triassic. Perhaps then there were two phases in the crisis: an initial time of thoroughly beastly conditions during which the Siberian Traps continued to erupt and greenhouse/anoxic conditions prevailed. The Chinese fossils show that life was not on track for recovery until 700,000 years after the crisis first hit. This was followed by an episode of 5 million years when plants on land were sparse, and forests had not become re-established, and when tetrapod communities consisted of generally small to medium-sized animals occupying a restricted range of niches, and not yet including larger herbivores or carnivores.

Our studies in Russia suggest a rather slow recovery of tetrapod faunas, with ecosystems seemingly still unbalanced at the end of the sampling period, some 15 million years after the mass extinction. The Middle Triassic ecosystems were again complex, but small fish-eaters and small insect-eaters were still absent, as were large herbivores and specialist top carnivores to feed on them. These gaps presumably reflect incomplete ecosystems and delayed recovery, rather than that the ecosystem had reached equilibrium at a lower level of complexity than is observed in the Late Permian. Evidence for this is that Late Triassic faunas from other parts of the world show fuller ecosystems – various amphibians as small fish-eaters, small insect-eaters, large herbivores, and large carnivores.

The same relaxed pattern of recovery seems to be true of plants and marine animals. For a long time, geologists had recognized a so-called 'coal gap' and 'reef gap' spanning the 20 million years of the Early and Middle Triassic. Vegetation on land was sparse, and trees in particular were rare, and in the seas, reefs had not re-established themselves. The end-Permian mass extinction had wiped forests and reefs from the Earth, and it took perhaps 20 million years for different species to re-invent these important modes of life.

So, at the level of ecosystems, the recovery lasted for 20 million years, much longer than the recovery period following the other big five mass extinctions. But those other events, even the extinction of the dinosaurs 65 million years ago, were far less severe, and major life modes were not so seriously devastated. The recovery after the end-Permian event took even longer on a global scale: counts of the number of genera of marine animals show that Late Permian totals were not achieved until the Late Jurassic, some 90 million years after the crisis, and families recovered their global richness at the end of the Jurassic, 100 million years after.

In the words of Leigh Van Valen, noted evolutionist and palaeontologist from the University of Chicago, the end-Permian mass extinction 'reset Phanerozoic community evolution'. The plants and animals after the event were different, and their modes of evolution changed too. For life in the sea and on land, the new world of the Triassic, the beginning of the Mesozoic Era, marked the beginning of the construction of modern ecosystems.

Chapter 7
The origin of modern ecosystems

> The dinosaur's eloquent lesson is that if some bigness is good, an
> overabundance of bigness is not necessarily better.
>
> Eric Johnston, President US Chamber of Commerce (1958)

It may seem counterintuitive to launch a chapter about the origin
of modern ecosystems with a quotation about dinosaurs. However,
it is true that modern ecosystems in the sea and on land began to
assemble themselves during the Triassic and, ironically it might
seem, dinosaurs were a part of those ecosystems.

The end-Permian mass extinction had been such a blow to life in
all settings that ecosystems took many tens of millions of years to
reconstruct themselves in the Mesozoic. The Permo-Triassic
boundary marks a profound division in the history of life, so
profound that it was noticed in the 1830s, long before
palaeontologists had a clear picture of the timing of major
events.

This chapter is about the Mesozoic Era, the time from 251 to
65 million years ago, marked at each end by a mass extinction
event, the end-Permian at the start, and the equally famous
end-Cretaceous, or Cretaceous-Tertiary (KT), event at the end that
saw the extinction of dinosaurs, marine reptiles, and ammonites.

The Mesozoic Era was named in 1840 by the polymath English geologist John Phillips (1800–74): he noticed dramatic differences between the fossils of the earliest rocks, which he assigned to the Palaeozoic ('ancient life'), the Cambrian to Permian periods inclusive, and those of the subsequent Triassic, Jurassic, and Cretaceous periods, which he termed collectively the Mesozoic ('middle life'). He assigned the remaining 65 million years from the end of the Cretaceous to the present day to the Kainozoic (sometimes Cenozoic, 'recent life').

In this chapter we will look at the key elements in the assembly of modern ecosystems, first in the sea and then on land. First, it is important to look at what the world was like through the Mesozoic.

The Mesozoic world

Setting aside the devastation of life that had taken place, the Triassic world was similar in many ways to the Permian. All continents remained united as the supercontinent Pangaea, although the North Atlantic Ocean began to open at the very end of the period, with rifting in eastern North America, southern Europe, and North Africa. Nevertheless, there is strong evidence that tetrapods could migrate widely because faunas of continental tetrapods were similar worldwide. For example, the first faunas of the earliest Triassic were dominated by the cosmopolitan plant-eater *Lystrosaurus* (see p. 111), as well as other small and medium-sized reptiles and amphibians that were more or less the same from place to place.

Triassic climates were warm, with much less variation from the poles to the equator than exists today. There is no evidence for polar ice caps, because the North and South Poles both lay over oceans at the time. During the Late Triassic, there was a broad climatic shift from warm and moist to hot and dry. This aridification may have been caused partly by a northwards drift of Pangaea so that critical areas moved into the equatorial zone. Whatever the causes, this climatic change may have been crucial in kick-starting modern ecosystems on land, and certainly in triggering the beginning of the age of dinosaurs.

Jurassic climates were moister than in the Triassic and warm conditions prevailed right to the poles. Ferns and conifers of subtropical varieties have been found as far north as 60 degrees palaeolatitude, and rich floras are known from Greenland and Antarctica.

Cretaceous climates were probably similarly warm, although there have been suggestions that ice caps existed at both poles during part of the Cretaceous. The floras show similar patterns to the Jurassic. Polar regions had warm-temperate climates and the boundary between the subtropical and temperate floras was 15 degrees closer to the poles than it is today. Thus most of the United States, Europe as far north as Denmark, and most of South America and Africa enjoyed tropical climates. Dinosaurs and other fossil reptiles are known from all climatic zones, from the equator to the poles.

During the Jurassic and Cretaceous, the Atlantic Ocean opened up progressively, unzipping from north to south. By the Late Jurassic, the North Atlantic was fairly wide, and dinosaurs could only just get across, perhaps via a northern bridge over Greenland. Africa was separated from Europe and Asia by oceans for much of the Jurassic and Cretaceous, and connected to South America. But that connection broke in the Early Cretaceous as the South Atlantic progressively opened. It seems that South America kept a narrow link round to Madagascar, and India and Australia moved rapidly eastwards in the Cretaceous, losing contact with Africa, and reaching their present positions after the end of the Cretaceous.

Life in Triassic seas

A scattering of marine organisms survived the end-Permian mass extinction – some brachiopods, molluscs, echinoderms, fishes, and reptiles. In the Early Triassic there was an episode of unusual conditions, and many marine organisms became rather tiny. In some locations, there are abundant fossils, but these may be all minute gastropods, coiled molluscs that are a quarter to a half of their normal pre-extinction size. This is called the Lilliput Effect. Richard Twitchett from the University of Plymouth has argued that the miniature gastropods, together with miniature fishes, sea stars, and others, all indicate a time of low food supplies. Perhaps the smaller species were the only ones that could survive in the post-extinction world, or perhaps individuals just evolved to be smaller.

Ammonoids, the coiled, swimming molluscs, were hard hit by each extinction event. Permian ammonoids had diversified considerably and occupied a range of ecological roles as free-swimming carnivores, feeding on plankton and small swimming crustaceans. They were nearly driven to extinction at the end of the Permian, but for two or three species that survived. These few re-radiated in the Triassic, and they had replaced most of their former ecological roles within 10 million years.

Corals had been devastated by the end-Permian mass extinction. The rugose and tabulate corals that had formed reefs throughout much of the Palaeozoic were gone, and there was a 'reef gap' of some 10 to 15 million years during the Early and Middle Triassic, during which there were no reefs. Then, the Scleractinia, the main modern group of corals, began to form small patch reefs in the Middle and Late Triassic, and they eventually took over the role that had been vacated, and constructed larger and larger tropical reefs, as we see today.

The end-Permian mass extinction had driven many fish groups to extinction, but the survivors diversified in the Triassic, including sharks of relatively modern appearance. The bony fishes radiated extensively, and most of the Triassic forms were less heavily armoured than their precursors. Some of these new fish groups became immensely diverse. Semionotids, for example, were small actively swimming fishes that occur in huge diversity in ancient lake systems down the eastern seaboard of North America. Some of their relatives were deep-bodied, others were long, and some had pointed snouts.

Triassic marine reptiles

The growing diversity of marine fishes in the Triassic, as well as all the other new seabed animals, and the ammonoids, provided a rich diet for larger predators. In the Permian, there had been few marine reptiles, but there was an astonishing diversification of such creatures in the Triassic (Fig. 19), some of them short-lived, others destined to be significant parts of the marine ecosystem for the whole of the Mesozoic.

Perhaps the most successful group of marine reptiles of the Mesozoic were the ichthyosaurs (literally, 'fish lizards'). Ichthyosaurs (Fig. 19A) were highly adapted for life in the sea, with their dolphin-like bodies – no neck, streamlined form, paddles, and fish-like tail. They arose in the Early Triassic and continued throughout the Mesozoic Era with essentially the same body form. Early ichthyosaurs, such as *Mixosaurus* from Germany, were 1 to 3 metres long, and with a long snout. The eyes were huge, and the jaws narrow and lined with uniform peg-like teeth. The fore- and hindlimbs were shortened and broadened as paddles, and the separate digits were almost certainly bound up in a 'mitten' of skin, as in dolphins today, so they could act most efficiently as propulsive organs. There is very little clue in the skeleton of *Mixosaurus* about its ancestry: ichthyosaurs clearly arose from land-living reptiles, but those ancestors are yet to be found.

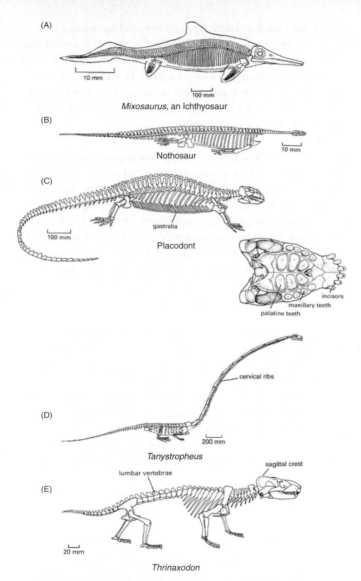

(A)

10 mm

100 mm

Mixosaurus, an Ichthyosaur

(B)

Nothosaur

10 mm

(C)

100 mm

gastralia

Placodont

incisors

maxillary teeth

palatine teeth

cervical ribs

(D)

200 mm

Tanystropheus

lumbar vertebrae

sagittal crest

(E)

20 mm

Thrinaxodon

19. Reptiles from the Triassic, including marine (A–D) and non-marine (E) forms.

Ichthyosaurs became relatively abundant, and hundreds of specimens are known from the Middle Triassic and Early Jurassic of England and Germany. In the Triassic, some ichthyosaurs from North America reached the huge length of 15 metres, although most remained in the 1- to 3-metre range. Ichthyosaurs bore live young, as attested by many remarkable skeletons of females that died just before, or during, childbirth.

Two groups of Triassic marine reptiles, the pachypleurosaurs and the nothosaurs, had long necks, and were not so streamlined and adapted to water life as the ichthyosaurs. Pachypleurosaurs were generally shorter than 50 centimetres, and they must have paddled around in pursuit of smaller prey, presumably fishes and crustaceans. The larger nothosaurs (Fig. 19B) could have tackled ichthyosaurs and larger fishes. These reptiles probably spent much of their time in the water, but might have come out on shore to lay eggs. Nothosaurs are closely related to the plesiosaurs, a group that rose to prominence in the Jurassic.

Another group of marine reptiles unique to the Triassic were the placodonts (Fig. 19C), whose name means 'pavement teeth'. This designation refers to their mouths, which were broad, and lined with some spatulate incisors at the front, and some massive flattened teeth on the palate and over the plate-like lower jaw. This array of teeth was clearly designed for crushing; the placodont dentition must have functioned like a pair of grindstones, and it is likely they fed on molluscs. Placodonts had heavily built, broad bodies, and large paddle-like limbs. They could have walked on the seabed, and swum through the water, and they presumably grazed on great oyster beds, scraping off shellfish with their incisors, and crushing the shells between their massive molars.

Strangest of all the marine reptiles was *Tanystropheus* (Fig. 19D), a land-liver that is often preserved in Middle and Late Triassic marine sediments side by side with pachypleurosaurs, ichthyosaurs, and placodonts. *Tanystropheus* had a hugely long neck, twice the length of the rest of its body. The neck was not greatly flexible because it was composed of only 9–12 cervical vertebrae. Juveniles of *Tanystropheus* have relatively short necks and, as they grew larger, the neck sprouted at a remarkable rate. Its function is a mystery. The sharp teeth suggest that *Tanystropheus* fed on meat, so it may have stood on rocks, or swum in shallow sea waters, and darted its neck around to catch fish.

By the end of the Triassic, 50 million years after the end-Permian mass extinction, reefs were established, and many superficially familiar kinds of shellfish, echinoderms, and others were moving around on the seabed. Some of the astonishing diversity of marine reptiles did not last beyond the Triassic, but others took their place in the Jurassic. Ecosystems on land had similarly recovered to a considerable extent during the Triassic, and a new reptile group had appeared, the dinosaurs.

Terrestrial ecosystems

During the earlier part of the Triassic, floras in the southern hemisphere were dominated by the seed-fern *Dicroidium*, a shrubby plant with broad leaves. Creeping through these plants were numerous millipedes, centipedes, spiders, and insects – none of them as huge as their Carboniferous precursors (see p. 93). Other denizens of the undergrowth included worms, snails, and slugs, to our eyes at least probably quite familiar-looking. These were preyed on by small reptiles, such as therocephalians and bauriids, surviving mammal-like reptiles from the Permian.

Shallow ponds and rivers were inhabited by a variety of amphibians, mainly temnospondyls, holdovers from the Carboniferous and Permian (see p. 97) that continued in this role until the Early Cretaceous. Triassic faunas were an interesting mixture of Palaeozoic survivors, such as the temnospondyls, and entirely new groups that presumably had their chance to evolve after the devastation of ecosystems by the end-Permian extinction.

Further small predators were the cynodonts (literally, 'dog teeth'), reptiles that looked quite different from the others of their day. An example is *Thrinaxodon* from the Early Triassic of South Africa (Fig. 19E). This animal probably looked somewhat dog-like, and may even have had hair – a seemingly dramatic assertion because hair is not fossilized. *Thrinaxodon* had a smallish head, jaws lined with differentiated teeth (incisors, canines, molars), a flexible neck, powerful back, and shortish tail. The limbs were actually held quite upright, tucked under the body, instead of sprawling out at the side, like most other animals in the Early Triassic, and it could perhaps have moved rather fast.

What about the suggestion of hair? If *Thrinaxodon* had hair, it was presumably warm-blooded, and so more like a mammal than a reptile. The evidence for hair is that *Thrinaxodon* has tiny pits on its snout, just like the openings around the snout of modern mammals through which the sensory nerves pass to supply the whiskers. Mammals use their whiskers for sensing their surroundings, and it seems that the early cynodonts also had whiskers. Whiskers are modified hairs, so if *Thrinaxodon* had whiskers, it also presumably had a pelt of hair over its whole body.

The Early Triassic cynodonts show further mammalian characters – the teeth are differentiated, unlike reptiles, which typically have more or less identical teeth from front to back of the jaws. In the backbone, too, *Thrinaxodon* shows two regions, the normal thoracic vertebrae at the front carrying ribs, and lumbar vertebrae behind, without ribs. Reptiles typically have ribs along the length of their thorax, while mammals have ribs around their lungs, then a strong diaphragm to pump the lungs, and no ribs around the belly region.

The rise of the ruling reptiles

The key beneficiaries of the end-Permian mass extinction on land were the cynodonts, such as *Thrinaxodon*, and the archosaurs, meaning literally 'ruling reptiles'. Archosaurs today include crocodilians and birds, but in the past the archosaurs were more diverse, and they include the extinct dinosaurs and pterosaurs. Archosaurs are distinguished from other reptiles by the possession of a so-called *antorbital fenestra*, an opening in the side of the skull between the nostril and the eye socket, of uncertain function. The first archosaur is known from the latest Permian of Russia.

In the earliest Triassic, archosaurs were modest-sized predators, up to 1 metre in length. But within 5 million years of the end-Permian extinction, *Erythrosuchus* was up to 5 metres long, and equipped with a massive skull, so that it was clearly capable of preying on anything. The archosaurs diversified enormously in the Middle and Late Triassic. Some became large carnivores, others became specialized fish-eaters, others adopted a specialized grubbing herbivorous lifestyle, yet others were small, two-limbed, fast-moving insect-eaters (the crocodilians and dinosaurs), and some took to the air (the pterosaurs).

The pterosaurs were proficient flapping flyers, with a lightweight body, narrow hatchet-shaped skull, and long narrow wings supported on a spectacularly elongated fourth finger of the hand. The bones of the arm and finger support a tough flexible membrane that could fold away when the animal was at rest, and stretch out for flight. Pterosaurs were covered with hair, and were almost certainly *endothermic*: that is, capable of generating heat within their bodies to maintain a high metabolic rate during flight.

The pterosaurs were important flying animals through the Jurassic and Cretaceous: mainly fish-eaters, but some were insect-eaters, and some might even have been scavengers on dinosaur carcasses, like giant vultures. Some later pterosaurs were much larger than any known bird, such as *Pteranodon* with a wingspan of 5 to 8 metres, and *Quetzalcoatlus* with a wingspan of 11 to 15 metres. Most pterosaurs fed on fishes caught in coastal seas, but others were insectivorous.

The first dinosaurs

Dinosaurs arose at the beginning of the Late Triassic, some 230 million years ago. The oldest examples are known from the Ischigualasto Formation of Argentina, *Eoraptor* and *Herrerasaurus*, both of them relatively well known from nearly complete specimens, and they give an insight into the days before the dinosaurs rose to prominence.

Eoraptor was a lightweight biped, about 1 metre long, and *Herrerasaurus* was larger, about 3 metres long. Both of these dinosaurs had long hindlimbs, adapted for fast running, and shorter arms equipped with strong hands for grasping prey. These dinosaurs also had major modifications in the hip region to improve their upright posture and speed of movement.

Eoraptor and *Herrerasaurus* lived side by side with larger and more abundant animals, such as the hefty rhynchosaur *Scaphonyx*, and the hippo-sized dicynodont *Ischigualastia*, a descendant of *Lystrosaurus*. This is clearly not an ecosystem where dinosaurs 'ruled', and yet later in the Triassic the whole landscape was full of dinosaurs.

Palaeontologists used to employ rather macho metaphors for the time of dinosaurs: the rule of the reptiles, dinosaur domination, the dinosaur dynasty. And, whereas dinosaurs at the end of their days were seen as rather washed up and tired out, at the start they were supposed to have been virile, active, dominant, and in most ways better than the beasts that preceded them. But can palaeontologists really talk in terms such as these?

The Carnian crisis

When I was a student, we were taught that the major groups of plants and animals had arisen, in succession, as improvements on what went before. The new group – here the dinosaurs – would prevail over the existing groups – here the rhynchosaurs and dicynodonts – by power of adaptation. After all, evolution is 'survival of the fittest', so it made sense that the succession of major life forms through time involved progression and improvement.

I remember questioning this assumption. It seems such an obvious idea at one level, but how can we be sure that the succeeding plants and animals are always 'better' than their precursors? Evolution is a process whereby individual organisms, and species, may improve their fit to the environment, but then environments keep evolving. So the target for adaptation is not static. Therefore, can we be sure that dinosaurs prevailed because they were wholly better than what went before?

As part of my Ph.D. work at the University of Newcastle, I looked at the data. It turned out that there was no evidence for a long-term replacement of primitive reptilian groups by the dinosaurs. The long-term competitive model might have suggested a succession of faunas in which dinosaurs became commoner and commoner, and their predecessors disappeared step by step. No such thing.

The critical time period is the Carnian Stage, from 230 to 220 million years ago. The Ischigualasto fauna of Argentina, and others of the same age around the world, contain the first smallish and rather rare dinosaurs. Then, a few million years later, in the succeeding Norian Stage, dinosaurs were everywhere. The rhynchosaurs, dicynodonts, and other herbivores had disappeared. Perhaps their disappearance was linked to the major climatic and floral change at this time, when climates went from humid to arid, and the *Dicroidium* flora was replaced by a conifer flora? Without the low, lush seed-ferns, rhynchosaurs and dicynodonts were scuppered. Only a new kind of herbivore could reach up into conifer trees, and cope with their spiky leaves.

There are three main divisions of dinosaurs, and all three groups arose at this time. The theropods, the flesh-eating forms, continued as mainly rather small animals, few of them exceeding a length of 3 metres in the Triassic. The limitation in size may be because there were still some other archosaur groups around, non-dinosaurs, that filled the ecological niche as top carnivores. These were the rauisuchians, hefty quadrupedal animals with 1-metre-long skulls and vast sharp teeth that could prey on the Carnian rhynchosaurs and dicynodonts, but also on the first herbivorous dinosaurs that replaced them.

The main beneficiaries of the end-Carnian extinction event were the sauropodomorphs, the second main line of dinosaurian evo-lution. These were animals like *Plateosaurus* in the Late Triassic, all herbivores, and ancestral to the giant long-necked dinosaurs of the Jurassic. *Plateosaurus* is known from dozens of skeletons from central Europe, from France to Greenland, and Holland to Poland, and it was clearly the dominant herbivore of its day.

Plateosaurus was up to 7 metres long, with a long neck, a massive body, and long tail. It was fundamentally bipedal, the primitive condition for dinosaurs, but it was becoming so large that it could stand and walk on all fours as well. It had a tiny head, and its jaws were lined with small, sharp, but rather leaf-shaped teeth, clearly adapted for plant-eating. Its hand had a vast, hook-shaped thumb claw that might have been used for sweeping in leaves and other plant material. Great herds of these animals roamed over Europe, and presumably the rest of the world: in North America, skeletons are not known, but many deposits are covered with their footprints.

The third dinosaur group, the ornithischians, are only sparsely known in the Late Triassic, but they rose to prominence in the Early Jurassic. The ornithischians were all herbivores, and they started as smallish bipeds, but soon radiated, in the Jurassic, to include unarmoured bipedal forms, as well as armoured quadrupeds. But are these really to be seen as parts of a modern ecosystem?

Modern faunas of the Late Triassic

The time-traveller who peered around in the Late Triassic might be forgiven for seeing the world as entirely alien. The plants included conifers, ferns, and mosses, typical forms today, but no flowering plants. The most obvious animals were dinosaurs of all sizes and pterosaurs in the air. But, a closer inspection might reveal a diversity of familiar forms.

Among the aquatic temnospondyls at the water side were the first frogs, and perhaps also salamanders. Further, the first turtles arose in the Late Triassic. As *Plateosaurus* stumbled around the central European forests, one or two heavily armoured turtles shuffled around the edges of lakes. Was the shell to protect the reptile from being stomped on accidentally by a dinosaur, or more likely to avoid predation?

There were also the first lizard-like animals. Modern lizards arose in the Jurassic, and snakes in the Cretaceous, but a basal group, the sphenodontids, originated in the Late Triassic. There is actually a living sphenodontid, the tuatara *Sphenodon* from New Zealand. This famous 'living fossil' lives in burrows and feeds on worms and insects. It has a heavier, more solid skull than true lizards, and many other primitive features, and probably is not much different from its Triassic forebears.

There were also the first crocodilians, which were largely terrestrial in habits, walked on all fours, and had an extensive armour of bony plates. Crocodilians became more diverse and abundant during the Jurassic and Cretaceous. Some even became fully marine in adaptations, to the extent that they had paddles instead of hands and feet, and a deep tail fin to speed their swimming.

Most striking of the 'modern' groups at this time were the first mammals. We have already seen that cynodonts like *Thrinaxodon* from the Early Triassic were warm-blooded, and had differentiated teeth, a diaphragm, and an upright, mammalian gait. By the Late Triassic one cynodont group had given rise to the first mammals. Little more than the size of shrews, these were nonetheless mammals, and they fed on insects, as attested by their needle-sharp little teeth. Fossils of these basal mammals are extremely rare, and they show how the reptilian skeleton became modified into a fully mammalian condition. The most astonishing change was that the reptilian jaw joint became embedded in the mammalian middle ear.

In reptiles the jaw hinge lies between the articular bone in the lower jaw and the quadrate bone in the skull. In mammals today, the jaw joint is between the dentary bone of the lower jaw and the squamosal bone in the skull. Through the Triassic, a remarkable series of fossils shows how the seemingly impossible happened: one jaw joint losing its function, and another joint being invented to take its place. Some Middle and Late Triassic cynodonts had *two* functioning jaw joints, but the reptilian one was becoming involved in hearing. Reptiles, like fishes, have a single auditory bone, the stapes, which is a simple rod that runs from the eardrum to the braincase. Mammals have three auditory ossicles, the famous hammer (malleus), anvil (incus), and stirrup (stapes). The stapes inside our ear is the original fishy hearing bone, while the articulated malleus and incus are the reptilian articular and quadrate. So our middle ear contains a remarkable reptilian remnant, a reminder of our evolutionary heritage.

Dinosaur evolution

The three major dinosaurian groups evolved and radiated substantially through the Jurassic and Cretaceous. Having already taken over the niches as medium to large herbivores and small carnivores in Late Triassic terrestrial ecosystems, they took over as large herbivores when the rauisuchians, and other archosaur groups, died out at the end of the Triassic.

Most striking of the Jurassic dinosaurs were the sauropods, such as *Brachiosaurus* and *Diplodocus*, known especially from the American Midwest. When these monsters were first discovered in the nineteenth century, many paleontologists thought that they were too big to have lived fully on land. It was assumed that the sauropods lived in lakes, supporting their bulk in the water, and feeding on waterside plants. New evidence shows, however, that life on land was quite possible, and indeed the long neck of *Brachiosaurus* made it a super-giraffe, a dinosaur that could feed on leaves from very tall trees, well out of the range of any other animal.

But how did sauropods become so large? *Brachiosaurus* and relatives reached lengths of 20 metres or more, and some weighed as much as 50 tonnes. An elephant reaches sexual maturity at about 15 years old, but the largest modern elephant is less than one-tenth the size of a large sauropod. Evidence from bone rings suggests that the sauropods actually grew very fast as juveniles, increasing their weight by 5 tonnes per year during a growth spurt between the ages of 5 and 12, by which point they had reached an adult weight of 25 to 30 tonnes. They could probably breed at this age, but would have continued growing larger, at a reduced rate, for several more years.

The sauropods apparently lived in herds, as suggested by trackways that show herds walking along ancient lake shorelines. They browsed on leaves from shrubs and trees, keeping their long necks more or less horizontal most of the time, but reaching up when necessary. Nests of sauropod eggs show that these dinosaurs scraped rough nests in the mud and laid up to a dozen rugby-ball-sized eggs on the ground. They may have covered the eggs with sand, and left them to hatch. It's not clear whether dinosaurs offered parental care to their young – it's a nice idea, often shown in animated films, but evidence is equivocal.

The first theropods were small, but the group diversified in the Jurassic and Cretaceous. The Late Jurassic *Allosaurus* was capable of hunting every other animal in the landscape, except for the sauropods, which were large enough, like elephants today, to escape most predation. Cretaceous theropods included *Deinonychus*, a human-sized, immensely agile and intelligent form – it had a bird-sized brain. Its key feature was a huge claw on its hind foot that it almost certainly used to slash at prey animals. *Tyrannosaurus* is famous as the largest land predator of all time, reaching a body length of 14 metres, and having a gape of nearly 1 metre.

Theropod dinosaurs gave rise to birds. Indeed, the oldest bird, *Archaeopteryx* from the Late Jurassic of southern Germany, has the skeleton of a small flesh-eating dinosaur, and yet it is covered with feathers. Many theropods had feathers – as shown by spectacular fossils from Liaoning in China – and those feathers were probably for insulation, to allow them to be warm-blooded. Some relatives of *Deinonychus* even had flight feathers along their arms and legs, and some may even have used these to enable them to glide down from a high point. *Archaeopteryx* was capable of powered flight, and birds evolved extensively through the Cretaceous, and then became particularly diverse and abundant after the end of the Cretaceous.

The ornithischians, the third major dinosaur group, diverged into armoured and unarmoured forms in the Jurassic (Fig. 20). Two groups of armoured ornithischians were the stegosaurs and the ankylosaurs. The stegosaur *Stegosaurus* has a row of bony plates along the middle of its back that may have had a temperature-control or display function. The ankylosaur *Euoplocephalus* is a massive tank-like animal with a solid armour of small plates of bone set in the skin over its back, tail, neck, and skull: it even had

20. Dinosaurs of the Late Jurassic of North America

a bony eyelid. The tail club was a useful defensive weapon that it used to whack threatening predators such as *Tyrannosaurus*.

Most ornithischians were ornithopods, bipedal forms, initially small, but later often large. In the Late Cretaceous, the hadrosaurs were successful, fast-moving, plant-eaters. Many of them have bizarre crests on top of their heads that may have been used for species-specific signalling, and their duck-billed jaws are lined by multiple rows of grinding teeth. Close relatives of the ornithopods were the ceratopsians ('horn-faces'), like *Centrosaurus*, which had a single long nose-horn and a great bony frill over the neck.

There has been a continuing debate about whether the dinosaurs were warm-blooded or not. Evidence for warm-bloodedness is strongest for the small active predators like *Deinonychus* that might have required the added stamina and speed. However, endothermy is costly in terms of the extra food required as fuel, and it is not clear whether the larger dinosaurs could have eaten fast enough. Indeed, larger dinosaurs would have maintained a fairly constant core body temperature simply because of their size, whether they were endothermic or not.

The Mesozoic Marine Revolution

In plots of the diversification of life in the sea or on land, there was a major kick upwards in diversity in the Cretaceous, some 100 million years ago. Diversity had slowly recovered after the end-Permian extinction and, as we have seen, ecosystems had more or less rebuilt themselves by the end of the Triassic. Many of the modern groups of organisms were around, but rapid diversification happened in the past 100 million years. This rise in diversity of the modern fauna and flora coincides with some remarkable events in the sea and on land.

It should be noted that this pattern of diversification has been questioned. Palaeontologists must rightly probe the quality of the fossil record: can they really be sure that the diversity of fossils collected in any way matches the diversity of life at the time? Whereas most palaeontologists agree that it would be wildly foolish to read every rise and fall of the diversification curve as a real event, the diversification kick from the Early Cretaceous onwards has been so sustained, and is repeated in every analysis, that it must be broadly real.

In the sea, there were three main aspects of the diversification pulse in the Cretaceous, sometimes termed the Mesozoic Marine Revolution: new kinds of plankton, changes in predation habits, and new vertebrate groups.

First, plankton groups diversified enormously through the Mesozoic. Coccoliths, simple plant-like organisms with calcium carbonate shells, arose in the Late Triassic, and became hugely abundant in the Cretaceous. They were so abundant, in fact, that their dead shells littered the sea floor and built up what were to become many hundreds of metres of chalk deposits in the Late Cretaceous of many parts of the world. Planktonic foraminifera also rose in the Late Triassic, and became hugely diverse and abundant in the Cretaceous, as did the dinoflagellates, organic-walled swimming algae, radiolaria, with their silica shells, and diatoms, single-celled brown algae.

Fiendish new modes of predation seem to have stimulated massive radiation of new groups of seafloor animals that cracked, crushed, and drilled into their prey. The ancestors of crabs and lobsters emerged in the Early Cretaceous, and they nipped and cracked shells and echinoderms. New kinds of shell-crushers, durophages, appeared, including the placodonts in the Triassic (see p. 128), as well as a broad array of shell-crushing fishes and reptiles from the Jurassic and Cretaceous. Some rather fiendish gastropods, essentially whelks, evolved extraordinary drilling capabilities. Shell-boring has long been an efficient mode of predation, but new groups in the Cretaceous refined this skill to new levels. The gastropod uses either chemical or physical means to cut a hole into the shell of its prey, and then slurps out the contents. It may secrete dilute acids that burn through the calcium carbonate shell, or use its toothed radula, a kind of tongue, to rasp a hole. Other predators hammered their prey against surfaces, speared them, sucked them out of their natural openings, swallowing them whole, or wrenched the flesh from the shell.

The prey organisms clearly had to increase their defences. Indeed what was happening was an 'arms race' in which predator and prey drove each other's evolution. As the attacks became more fiendish, shellfish and echinoderms evolved thicker shells, additional armour, or escape mechanisms. Squid for example squirt a dense cloud of blue-black dye into the water if they feel threatened, and swim backwards by rapid blasts of water through a siphon just below their heads. Ammonites may have behaved similarly. This would confuse most predatory fishes and reptiles.

Seabed creatures often became effective burrowers, and there had never been so much burrowing as in the Cretaceous.

Modern fishes evolved rapidly in the Jurassic and Cretaceous. Neoselachians, the modern sharks, appeared in the Triassic, but became diverse and abundant especially in the Cretaceous. Their evolution paralleled that of their principal prey, the bony fishes. The major group of bony fishes today are the teleosts, some 23,000 species, ranging from salmon to goldfish, and seahorse to plaice. Teleosts are not armoured, but rely on fast movement, and a flickering, silvery sheen that confuses their predators. Unlike earlier kinds of bony fishes that had rather simple mouths, teleosts can project their mouth in a remarkable 'pout' that enables them to suck up their prey, and they have adapted their jaws to burrowing, snipping off coral, and many other functions.

The top predators in Jurassic and Cretaceous seas were the marine reptiles – ichthyosaurs and plesiosaurs that evolved from Triassic ancestors, as well as an extraordinary Cretaceous group, the mosasaurs. Mosasaurs were lizards that had secondarily become marine in habits, and reached huge size, some of them 10 metres long. They were significant predators on fishes and ammonites in the Late Cretaceous, but, together with the ichthyosaurs, plesiosaurs, and dinosaurs, all became extinct by the end of the Cretaceous.

The Cretaceous terrestrial explosion

There was an equally important explosion of life on land from the Cretaceous onwards. As we have seen, many modern groups of tetrapods had become established by the Late Triassic – frogs, turtles, crocodilians, lizards, and mammals. But the plants and insects were still of more primitive kinds. This all changed in the Early Cretaceous.

Triassic and Jurassic landscapes contained low ferns, horsetails and cycads, and tree-sized clubmosses, seed-ferns, and conifers. In the Early Cretaceous, the first flowering plants (angiosperms) appeared, and they radiated rapidly during the Late Cretaceous until they reached modern levels of diversity. The earliest angiosperms include magnolia, beech, fig, willow, palm, and other familiar flowering shrubs and trees.

The angiosperms are by far the most successful plants today, with over 260,000 species and occupying most habitats on land. The angiosperm flower on its own was not the reason for the success of the angiosperms, but rather their 'double fertilization' mode of reproduction. The sperm travels in the form of pollen, and fertilizes the ovule in a female flower. At the same time, additional sperm 'fertilize' another part of the female plant tissue that produces a nutritive tissue for the developing embryo. This offers economies over other modes of plant reproduction in which food stores in the seeds may be wasted if the embryo is not fertilized.

Pollen must be transported efficiently, whether by wind or water (as in more primitive plants) or, better, by insect power. Many flowers have coevolved with particular insects, bats, or birds so that the pollinator is obliged to work hard on behalf of the angiosperm. By means of a nectar reward, particular pollinating animals are drawn to particular species of flower, they enter to drink some nectar, and they move on to the next flower, carrying pollen that is then passed to the female plant structures where fertilization may then take place. Flowers, as has often been said, are the plant's way of enslaving bees, moths, bats, and other pollinating animals.

The rise of angiosperms in the Cretaceous drove a major radiation of insects at the same time. Groups of beetles and flies that pollinate various plants were already present in the Jurassic and Early Cretaceous, but the hugely successful butterflies, moths, bees, and wasps are known as fossils only from the Cretaceous and

Tertiary. The evolution of particular families of bees and wasps, for example, can be tied closely to the evolution of particular angiosperm groups.

Extinction

The Mesozoic is commonly called the 'age of dinosaurs', and dinosaurs would certainly have seemed very obvious in the landscape had an intrepid time-traveller gone back to the Jurassic or Cretaceous. In many ways, however, the dinosaurs were an unusual, admittedly large, sideshow that swamped from view many of the more important aspects of the evolution of marine and terrestrial ecosystems.

The mass extinction at the end of the Cretaceous, commonly called the KT event ('K' for Cretaceous, because 'C' was already assigned to the Carboniferous; and 'T' for Tertiary), is one of the big five mass extinctions (see p. 102). It has been studied in huge detail, but we will just give a bare outline of what happened.

Although much debated over the years, the KT event was almost certainly caused by a massive meteorite impact on the Earth. There is indeed evidence for long-term cooling of the climate, and for massive volcanic eruptions in India (the Deccan Traps), but these on their own do not seem to have caused the mass extinction.

The meteorite, some 10 kilometres across, struck the Earth in the region of the Yucatán Peninsula, in southern Mexico. The impact created the huge Chicxulub Crater, some 150 kilometres across, and now entirely covered by more recent sediments. Geophysical maps and boreholes through the structure show that the crater extended to the Earth's mantle before it rebounded. There are melt rocks in the floor of the crater, and the impact sent great tsunamis, tidal waves, sideways across the proto-Caribbean, which beat the shores of the Americas and shifted boulders as big as houses.

The exploding meteorite also sent vast clouds of dust into the upper atmosphere, and this settled out over the course of many months across the surface of the Earth, marked by the rare element iridium, which came from the core of the meteorite. These clouds blacked out the Sun and caused global freezing and a cessation of photosynthesis by land plants and by plankton. This cut away the basis of many food chains and led to widespread extinction. What the darkness did not kill succumbed to the icy cold of a dark world.

The large reptiles all disappeared – dinosaurs, pterosaurs, ichthyosaurs, plesiosaurs, mosasaurs – but frogs, salamanders, lizards, snakes, turtles, crocodilians, birds, and mammals all survived, with some major losses here and there. Plants and insects were only affected temporarily. Among sea creatures, the ammonites and other Cretaceous forms died out, but most animals on the seabed survived. Foraminifera were hard hit, as were other elements of the plankton.

The KT event was a serious extinction that deeply affected the evolution of life, and especially, it could be argued, cleared some space in terrestrial ecosystems for the distant ancestors of humans to populate.

Chapter 8
The origin of humans

Why was man created on the last day? So that he can be told, when pride possesses him: God created the gnat before thee.

The Talmud

The universe may have a purpose, but nothing we know suggests that, if so, this purpose has any similarity to ours.

Bertrand Russell

Why should the final chapter of this book be about the origin of humans? The argument could be that this falls in correct chronological order: we have looked at the origin of life, sex, skeletons, land life, dinosaurs, and then humans come next. That's all very well, but we could equally look at the origin of sparrows or cats or sweet potatoes. It is virtually impossible not to focus on human origins, because we are human. Therein lies a danger. Humans are not the pinnacle of evolution. Everything that has gone before was not a prelude to the appearance of human beings who arrived to a great fanfare.

But humans are special: no other species on Earth, to our knowledge, writes books, or even reflects on the history of its own species. Wise philosophers through the ages have warned us to be humble. But humility is not what this is about. The key point is that evolution is not teleological, or 'goal driven'. There can be no pathway laid out into the future for evolution: species come and go, buffeted by the vicissitudes of history.

In the Late Triassic, a naturalist would have had no reason to suppose that dinosaurs would become large and diverse, and dominate terrestrial ecosystems for more than 160 million years, and that mammals would remain small and humble denizens of the night. Equally, when the dinosaurs were cleared from the surface of the Earth by the KT mass extinction, crocodilians, birds, or mammals all stood a reasonable chance of becoming top predators. In South America, which was isolated from other parts of the world, certain crocodilians became rather terrestrialized, and took on the role of predators. In South America too, but also in North America and Europe, giant birds with massive 1-metre long bone-cracking beaks preyed on the ancestors of horses and cats.

At the beginning of the new world, during the Palaeocene Epoch of the Tertiary Period of the Cenozoic Era (see p. 18), the closest ancestors of humans were weedy little squirrel-like animals scuttling nervously along the tree branches. No sign there of latent genius or incipient domination of the Earth.

The first primates

An intriguing fossil was reported in 1965 as 'the oldest primate'. The specimen was named *Purgatorius*, after the locality Purgatory Hill in Montana near where it was found, and the report caused a sensation. Here was our distant ancestor, living side by side with *Tyrannosaurus rex* and peering, perhaps knowingly, from behind a branch. Sadly, this report has subsequently been discounted – the fossil was just an isolated tooth, but the identity of the tooth is not so much in question as the age of the rocks in which it was found. The tooth was in a channel that had been cut down into the latest Cretaceous from the overlying Palaeocene, and so post-dated the dinosaurs. Since 1965, no other convincing find of a Cretaceous primate has been reported, and the image of monkeys and dinosaurs living together cannot be confirmed – even though molecular evidence strongly points in that direction. More of that anon.

The primates are one of the eighteen orders of modern placental mammals, named from the Latin *primus*, 'first'. As primates, it was our privilege to call ourselves members of the 'first' order – this privilege extended to the Church as well, in which bishops and archbishops are termed primates. There was a time when books entitled 'The sex life of primates' could not be sold safely in England. All primates share a number of features that give them agility in the trees (mobile shoulder joint, grasping hands and feet, sensitive finger pads), a larger than average brain, good binocular vision, and enhanced parental care (one baby at a time, long time in the womb, long period of parental care, delayed sexual maturity, long lifespan).

Purgatorius is a plesiadapiform, a group of squirrel-like animals that may have climbed trees. They had long tails, grasping hands, and fed on fruit and leaves.

The plesiadapiforms lived among a diverse array of unusual mammals in the 10 million years following the KT extinction. Mammals had of course originated much earlier, in the Late Triassic (see p. 136), and they had diversified substantially during the Jurassic and Cretaceous, but most of the Mesozoic groups either died out during the Mesozoic or soon after. The three modern orders originated in the Jurassic and Cretaceous as well – the monotremes, marsupials, and placentals.

Monotremes today are restricted to Australia and New Guinea, being represented by the platypus and the echidnas. These mammals are unique in still laying eggs, as the cynodont ancestors of mammals presumably did. The young hatch out as tiny helpless creatures, and feed on their mother's milk until they are large enough to live independently.

Marsupials, such as the modern kangaroos, koalas, and wombats of Australia and the opossums of the Americas, are reproductive intermediates. They do not lay eggs and produce live young, but these young are tiny and underdeveloped, and they complete their gestation in the mother's pouch.

Placental mammals are the most diverse of the three living groups. They produce young that are retained in the mother's womb much longer than is the case in marsupials, and they are nourished by blood passed through the placenta. Placental mammals are remarkable for their diversity of sizes, from tiny shrews and bats weighing a few grams, to the African elephant, weighing up to 5 tonnes, and the blue whale, weighing perhaps 100 tonnes (although no one has ever weighed a large whale, nor does anyone know quite how this could be achieved). Their ecological and geographic range is vast too, from desert-living rodents to polar bears, and from bats to whales.

Mammals, morphology, and molecules

Mammalogists have struggled for two or more centuries to understand the relationships of the major groups of living placentals – are cattle related to horses, bats to monkeys, whales to seals? Some morphological evidence was found to show that, for example, rabbits and rodents are closest sisters, elephants are closely related to the enigmatic African hyraxes and the aquatic sirenians, but many other supposed relationships were hotly disputed. Ironically, the more effort palaeontologists applied to this question, the less certain were their conclusions.

Molecular phylogeny reconstruction methods seem to have cut through the knotty problem. The story began in 1997, when Mark Springer of the University of California Riverside and colleagues discovered the Afrotheria, a *clade* (a clade is a group in an evolutionary tree that originated from one ancestor and includes all descendants of that ancestor) consisting of African animals, linking the elephants (Proboscidea), hyraxes, and sirenians with the aardvarks (Tubulidentata), tenrecs, and golden moles. The last three groups had all been assigned various positions in the classification of mammals, but their genes show they shared a common ancestor with the elephant–hyrax–sirenian group.

After 1997, everything else seemed to fall into place. The South American placentals, the edentates, formed a second major group, the Xenarthra. And the remaining mammalian orders formed a third major clade, the Boreoeutheria ('northern mammals'), split into Laurasiatheria (insectivores, bats, artiodactyls, whales, perissodactyls, carnivores) and Euarchontoglires (primates, rodents, rabbits). So, in the course of two or three years, several independent teams of molecular biologists solved one of the outstanding puzzles in the tree of life.

But why had this proved to be such a phylogenetic puzzle? Some suggest that the major splits among placental mammals happened very rapidly, and there was no time for shared morphological characters to become fixed. But the morphologists are fired up to find such characters: if the Afrotheria really is a clade, then there must be some obscure anatomical feature shared among them all! Early suggestions included the prehensile snout (elephants, tenrecs) or testicondy (retention of the testicles within the abdominal cavity). But none of these really applies to all afrotheres. In 2007, a possible shared morphological character was at last identified: afrotheres all have additional vertebrae in the lower back.

The other element of the debate had been the timing of all these splits. Early molecular analyses, around 1995, yielded dates for the deep divergences of placental mammals at 120 to 100 million years ago, well within the Early Cretaceous. Although placental mammals of that age are known, and these include the spectacular little *Eomaia* from China, these early fossil forms do not belong to any of the modern orders or superorders. The molecular dates were a challenge to palaeontologists, because the oldest fossils belonging to modern placental orders came from after the KT mass extinction, as in the case of the plesiadapiforms like *Purgatorius*.

Many palaeontologists, including myself, argued that the molecular dates for placental mammal origins must be too ancient, and perhaps for similar reasons of mis-calibration as the initial ancient dates for metazoan radiation in the Precambrian (see pp. 67–8). Indeed, some of the dates were revised upwards, but only into the 100- to 90-million-year range. Additional solace came from the zalambdalestids and zhelestids, fossils found in the mid Cretaceous, perhaps 90 million years ago, of Uzbekistan. These fossils were assigned to basal positions among the Boreoeutheria, and so seemed to fill substantial gaps in the fossil record.

Peace appeared to be about to break out, until a bombshell struck in 2007. In a thorough re-examination of the zalambdalestids and the zhelestids, as well as other Cretaceous placental fossils, John Wible from the Carnegie Museum in Pittsburgh, and colleagues, showed that they all fell outside the clade of modern placentals. So, the gap was restored: molecular data suggest clearly that there had been a considerable amount of placental evolution in the Late Cretaceous, with the basal split into South American, African, and northern clades, and their subsequent division into the major orders, including primates. The oldest fossils are unequivocally Palaeocene and Eocene in age, so there is at least a gap of 25 to 30 million years where fossils are seemingly absent.

It has been easy for supporters of the ancient molecular dates to say that the Late Cretaceous fossil record is deficient. It is true that mammal fossils are rare, but the point is that several dozen species of mammals are known from a number of Late Cretaceous localities around the world and yet, despite Herculean efforts to assign these to modern orders and superorders, such efforts have been rejected by Wible and colleagues. If primitive placentals are being found, and sometimes as quite complete fossils, where are the missing modern forms? This debate is likely to rumble on for a while.

The monkey-rabbits

One of the surprises of the molecular phylogenies was that primates were allied with rodents and rabbits. The molecular studies confirmed that primates are members of Archonta, the clade comprising also Scandentia and Dermoptera. The Scandentia are the tree shrews, a tiny order of some nineteen species of tree-climbers from south-east Asia. The Dermoptera, or flying squirrels, are only two species, both of which have a skin membrane between their arms and legs, down each side of the body, and they can glide from tree to tree. The three orders within Archonta are all characterized by some shared features of the ear region of the skull, as well as by 'the possession of a pendulous penis suspended by a reduced sheath between the genital pouch and the abdomen'.

Another clade that had long been recognized by the morphologists was the Glires, consisting of rodents and rabbits. The Order Rodentia is by far the largest placental order, comprising 2,000 species, about 40 per cent of all mammals. Rodents are fiendishly adaptable, and rats and mice have proved highly successful in human environments. The group includes also the cavies of South America, some of them quite large, as well as squirrels, beavers, and porcupines. Rabbit and hares, order Lagomorpha, share with rodents their constantly growing incisor teeth, a key factor in the success of both groups.

The molecular evidence confirmed the reality of the clades Archonta and Glires, and that both were close relatives within a larger clade, termed, with true inventiveness, but without regard for our dentures, Euarchontoglires. The Euarchontoglires are a major subdivision of the northern superorder of mammals, the Boreoeutheria, and so we must look to the northern hemisphere for the origin of those groups. Indeed, the oldest fossil primates and rodents, for example, come from the Palaeocene of North America and Europe.

Basal primates

Living primates are sometimes divided into prosimians, monkeys, and apes. These are convenient enough terms, although the 'prosimians' include quite a ragbag of forms that are neither monkeys nor apes, such as lemurs, lorises, and tarsiers.

There are over fifty living species of lemurs, which include the lemurs, indrises, and the aye-aye, all of them restricted to the island of Madagascar. Lemurs have long bushy tails, often striped black and white, and they are nocturnal, feeding on insects, small vertebrates, and fruit. The indrises include the woolly lemur, which is nocturnal and lives in trees, while the indri and the sifaka are diurnal animals that live in troops on the ground, and rarely move about bipedally by leaping along the ground. The aye-aye (*Daubentonia*) is a cat-sized nocturnal animal that probes for insects in tree bark with its slender elongated fingers.

Close relatives are the lorisiforms, thirty-two species of lorises and galagos (bushbaby), known from Africa and southern Asia. Fossil lemurs were known, until recently, only from Madagascar, but an earlier possible relative has been found recently in Pakistan, and the oldest possible loris fossil comes from the Eocene of Egypt.

Tarsiers, two species from the Philippines and the Indonesian islands, are tiny animals with huge goggly eyes that live furtively in the trees, and feed on insects, snakes, and birds. A diversity of ancestors of the tarsiers, as well as the entirely extinct omomyids and adapids, were significant tree-living animals in the Eocene of

North America and Europe, but also, later, in Africa and Asia. What other mammals lived at that time?

Tiny horses and giant rhinos

Most people have images of the evolution of horses, and some of the other familiar mammal groups. The first 10 million years of the Cenozoic saw a great deal of experimentation among mammals. The eighteen modern orders diversified, as well as a number of groups that have since become extinct. So it seems there was a kind of sorting out of the major mammalian lineages in the Eocene, which spanned from 56 to 34 million years ago.

Eocene horses, such as *Hyracotherium*, were indeed tiny, no larger than a terrier. *Hyracotherium* was a secretive woodland-dweller, adapted to scuttle through the tropical forests of Europe and North America, feeding on succulent leaves from the trees. The ancestors of cattle and of the flesh-eating carnivores such as lions and bears were also smallish woodland-dwellers.

Then a major habitat change occurred in the Oligocene, some 34 to 23 million years ago. Climates had been becoming slowly cooler since the end of the Mesozoic, and this cooling caused the climates in the centres of the continents to become arid. The lack of moisture meant that the lush forests died back, and grasslands spread more and more. Grasses had originated in the Cretaceous, but they did not become a dominant group in world ecosystems until the Oligocene. The secretive, camouflaged, forest mammals were squeezed into smaller and smaller patches, and many of them went extinct. Others ventured out onto the new savannas, and set off on a new evolutionary course.

The plant-eaters, such as the ancestors of cattle and horses, evolved to be larger. They had had four or five toes on each leg, and these reduced to three, and then to two in cattle and one in horses. The toe reduction was part of a process of leg lengthening and adaptation to fast running. On the open plains, camouflage was no longer a useful means of escape from predation, but height and speed were advantageous. As horses and cattle evolved to be larger, their teeth changed too. Leaves are relatively soft, but grass is hard because it contains small grains of silica. Horses and cattle evolved deeply rooted, ever-growing teeth with complex ridges of enamel and dentine on top to assist in grinding their food.

Some plant-eaters became huge. Among the afrotheres, elephants in Africa evolved from the size of pigs or small hippos to their modern size, and their trunks descended so they could continue to reach the ground. Rhinos, relatives of the horses, evolved into a diversity of forms, from medium-sized to very large. The largest of all, *Indricotherium*, was 5 metres tall, and looked like a cross between a buffalo and a giraffe.

As their prey increased in size and speed, the predators had to adapt too. Bears stayed largely in the woods, and continued to hunt woodland creatures, but also diversified their diets to include fruit, honey, and fish. Dogs never became very large, but they adopted new social structures, hunting much larger prey in packs, and relying on their endurance and their intelligence, to harry their prey to death. Some cats, such as lions and tigers, became large, and used their stealth to be able to creep up on their prey unseen, and then make a mad dash at the last minute.

Other mammal groups adapted in their own ways. Bats and whales were committed to life in the air and in the oceans respectively. By emerging at night, bats had found a new set of niches that birds did not occupy. Whales, descendants of land-living creatures, rediscovered the role of giant marine predators, vacated at the KT mass extinction by plesiosaurs and mosasaurs. The mammals of South America and Australia evolved rather independently from those of the Old World: Australia became a land of marsupials, including giant kangaroos and wombats, and South America had its own unique groups that mimicked horses, cattle, and rhinos.

Africa remained an island for most of the Cenozoic, but land bridges were formed across the Arabian Peninsula from time to time so that elephants and primates were able to pass across into Asia. Into this land of expanding savannas and diminishing forests came the first monkeys and apes.

Monkeys

After the Eocene, the omomyids and adapids became extinct, and modern 'prosimians' survived in some obscurity in Madagascar and south-east Asia. But a major new clade, the Anthropoidea, or monkeys, had arisen and was growing in importance. Monkeys differ from their precursors in having rounded, instead of the slit-like, nostrils, large canine teeth, and premolars and molars, the cheek teeth, modified for crushing plant material.

The origin of anthropoids is hotly debated: the traditional view is that the clade originated in Africa, while a new proposal is that they arose in Asia. The oldest African anthropoid appears to be *Algeripithecus* from the Middle Eocene of Algeria, based on isolated molars. More complete materials of anthropoids are known from the late Eocene of Egypt, and some of these show remarkable sexual dimorphism (physical differences between the sexes), as in many modern monkeys, where the males were twice the size of the females. This suggests that there was already a pronounced social structure, with males perhaps fighting their rivals for control of substantial harems of females. The Asiatic primates include several forms from the Eocene of China and Thailand: some may be non-monkeys, perhaps adapids, but others appear to be true anthropoids, and so further work is required to establish which came first, the monkeys from Africa or from Asia.

Monkeys today are divided into catarrhines, the Old World monkeys, and platyrrhines, the New World monkeys. These two groups seem to have diverged back in the Eocene or Oligocene, when the first platyrrhines somehow floated or swam across from Africa to South America. The two groups may be distinguished by their nose shapes: catarrhines ('narrow noses') have narrow noses with the nostrils placed below the nose, while platyrrhines ('broad noses') have broad noses and forward-facing nostrils. Platyrrhines also have prehensile tails, so if you see a monkey hanging by its tail, it's from South America.

New World monkeys include capuchins, tamarins, marmosets, and howler and spider monkeys. The Old World monkeys are more diverse, with smaller, tree-dwelling colobus monkeys, and larger ground-dwellers, such as baboons and mandrills, well known from Africa, but including also the Barbary ape of Gibraltar. These ground-dwellers live in large troops, the males are often much larger than the females, and they have often reduced or lost their tails. It's no wonder perhaps to realize that a specialized group of Old World monkeys became the apes.

Apes

The apes arose from the Old World monkeys before the end of the Oligocene and the group radiated in Africa in the Miocene. Indeed, Africa in the Miocene, 23 to 5 million years ago, has sometimes been described as the continent of the apes. A typical early form is *Proconsul*, which was named in 1933 on the basis of some jaws and teeth from Kenya. The name refers to a chimp called Consul who then lived at London Zoo, and entertained visitors with his bicycle riding and pipe smoking. Since the 1930s, much of the skeleton of several specimens of *Proconsul* has been found, and these show that this earliest of apes had a monkey-like body, and it probably ran along branches, and fed on fruit.

Proconsul was clearly an ape, and not a monkey, because it had no tail, its brain was relatively large, and it had broad molar teeth with a particular arrangement of the cusps, just as in the modern apes and in humans. Early ape evolution happened in Africa, but there were several migrations out of Africa between 25 and 10 million years ago. One migration passed through the Middle East to Europe, and Late Miocene apes are known from Hungary to Spain. Other migrations passed eastwards into the Indian subcontinent and south-east Asia.

The Asiatic ape diaspora included the ancestors of modern gibbons and of orang-utans, which diverged from the African ape line about 25 and 20 million years ago respectively. The gibbons have an obscure fossil history, but they specialized in *brachiation*, swinging through the trees from arm to arm. The orang-utan also evolved in south-east Asia, and specialized in tree life. Orangs evolved through a number of fossil forms called collectively the ramamorphs. The oldest examples come from Africa, but later ramamorphs, such as *Ramapithecus* and *Sivapithecus*, are known from Turkey, Pakistan, India, and China. *Sivapithecus* was rather like the modern orang-utan, with heavy jaws and broad cheek teeth covered with thick enamel, all of which suggest a diet of tough vegetation. The most extraordinary ramamorph was *Gigantopithecus*, which was ten times the size of *Sivapithecus*, and adult males might have reached heights of 2.5 metres and weights of 270 kg. This huge animal stalked the forests of south-east Asia from 5 to 1 million years ago, and many regard it as the source of stories of yetis in central Asia, and Big Foot in North America.

The gorillas and chimpanzees continued to evolve in Africa. Although both groups are able to walk on the ground using their own peculiar mode of locomotion, knuckle walking, they prefer to remain deep in the forests, and move slowly about in the trees, feeding on fruit and leaves. It has always been clear that gorillas and chimpanzees are the closest relatives of humans, but the fossil record is not helpful in providing evidence. However, as is well known, DNA evidence suggests that humans share most of their genome with the African apes, and our similarity to chimpanzees is greatest. The current best estimates from molecular clocks and from fossils are that gorillas diverged first, about 10 million years ago, and the ancestors of humans and chimps separated about 6–8 million years ago.

What is a human?

Early palaeoanthropologists followed the understandable prejudices of their time and assumed that the first humans must have distinguished themselves from their ape relatives by possession of a large brain. There was logic in this: among modern humans, our large brain could be said to be our defining characteristic. Whereas the brain volume of modern humans is 1,200–1,400 cubic centimetres (cc), gorillas have a 500 cc brain volume, and chimps a meagre 350 cc.

Brain volume gives a rough measure of intelligence, but only when considered in proportion to body mass. (Note that the blue whale has the largest brain volume of all, some 9,000 cc, but we would perhaps be loath to say that whales are eight times as intelligent as humans.) What matters is the *encephalization quotient* (EQ), the ratio of brain volume to body mass. The EQ for a whale is 1.8, higher than that for a horse (EQ = 0.9) or a cow (EQ = 0.5). Apes of course have rather high encephalization quotients, and the values are, in ascending order: gorilla (EQ = 1.6), chimpanzee (EQ = 2.3), and human (EQ = 7.5).

The other key human characteristic is bipedalism, walking on our hind legs. Dinosaurs and birds evolved bipedalism independently, and some lizards, monkeys, and apes can dash about on their hindlimbs for short spells. Among mammals in general, and primates in particular, humans are the only accomplished bipeds. Standing and walking fully upright led to many profound anatomical changes in our skeletons: the foot became flat and ceased to be able to grasp things, the ankle and knee became rather simplified hinge-like joints, and the hip joint modified enormously so the thigh bone fits into the hip socket with an inturned head. The pelvis has become bowl-like to support the guts, and the backbone is held more or less vertically, and it is S-shaped to accommodate the new pressures exerted by gravity. Quadrupedal mammals, including gorillas and chimps, have a long pelvis and a massive rib cage to hold the guts.

All the other peculiarly human characteristics stem from these two features. The large brain permitted or enabled language, social groups, extended care of children, adaptability to challenging environments, and technology. Bipedalism freed the hands for gathering food, tool-making, pot-making, scratching, and writing.

It seemed clear that humans acquired their large brains first, and then bipedalism. Early fossil discoveries in the nineteenth century, such as Neanderthal man from Germany and Java man, *Homo erectus* from Java, did not help much because palaeontologists were unsure of their relative ages.

The key support for the 'brain-first' theory came in 1912 when a remarkable skull was found in southern England, at the village of Piltdown. Here was an early human with a large brain. When the first important finds were reported from Africa in the 1920s, their significance was not realized, and it was only when Piltdown man was shown to be a fake in the 1950s that the true story emerged.

The skeletons of early hominids from Africa showed that bipedalism had arisen by 4 to 6 million years ago, and yet the increase in brain size came much later, perhaps 2 to 1 million years ago. Perhaps the first humans were forced to become bipeds as the central African forests diminished in size and the grasslands expanded between 10 and 5 million years ago. Modern chimps and gorillas are restricted to the great Congo forests in the west, whereas the first human fossils are known from a broad crescent over East Africa from South Africa, up through Kenya, Tanzania, and Ethiopia, to Chad in the middle of the Sahara Desert.

Sacré bleu! Les fossiles humains les plus vieux – ou non?

Until 2000, the oldest human fossils had been reported from rocks dated in the range 4 to 2 million years. Then, in 2001 and 2002, two rival French teams reported much older human fossils, each about 6 million years old. Both finds proved controversial, and there has been much name-calling and squabbling over the respective finds.

First was the report by Brigitte Senut and her team from Paris. In 2001 they reported the new hominid *Orrorin tugenensis* based on teeth, jaw fragments, and limb bones from Kenya. Senut and her colleagues argued that the teeth were rather ape-like, and that the arm bones suggested *Orrorin* could brachiate like an ape. However, the femur showed that *Orrorin* stood upright, and so this was a true early human.

The second discovery was by Michel Brunet and his team from Poitiers who reported *Sahelanthropus* from Chad in 2002. *Sahelanthropus* is based on a rather complete skull, some fragmentary lower jaws, and teeth. The *Sahelanthropus* skull indicates a brain volume of 320–80 cc, similar to a modern chimpanzee, but the teeth are more human-like, with small canines. The position of the foramen magnum, the hole through which the spinal cord passes out of the brain, is disputed: Brunet claims it is located beneath the skull, which would indicate that *Sahelanthropus* stood upright.

The australopithecines

The oldest substantial hominid skeletons, *Praeanthropus afarensis*, come from rocks dated at about 3.2 million years ago, and these show clear anatomical evidence for advanced bipedalism, but still an ape-sized brain. The famous skeleton of a female *P. afarensis* from Ethiopia, called Lucy by its discoverer Don Johanson in the 1970s, has a rather modern human pelvis and hindlimb. The pelvis is short and horizontal, rather than long and vertical as in apes, the thighbone slopes in towards the knee, and the toes can no longer be used for grasping. Lucy's brain, however, is small, only 415 cc for a height of 1 to 1.2 metres, and this yields an encephalization quotient not much different from a chimpanzee.

The human genus *Australopithecus* continued to evolve in Africa from about 3 to 1.4 million years ago, giving rise to further small species, including *A. africanus*, the species Raymond Dart first found in 1924. These australopithecines show advances over *Praeanthropus afarensis* in the flattening of the face and the small canine teeth. They also show some specializations that place them off the line to modern humans. For example, the cheek teeth are more massive than in *A. africanus* or modern humans, and they are covered with layers of thick enamel, adaptations to a diet of tough plant food.

The robust australopithecines, sometimes called *Paranthropus*, reached heights of 1.75 metres, but their brain capacities did not exceed 550 cc, still a rather ape-like measure. They had broad faces, huge molar teeth, and a heavy sagittal crest over the top of the skull, a feature also seen in large male gorillas. These are all adaptations for powerful chewing of tough plant food. Even the sagittal crest supports this interpretation since it marks the upper limit of jaw muscles that were much larger than in *A. africanus* or in *Homo*. The robust australopithecines may have fed on tough roots and tubers, while the gracile *A. africanus* perhaps specialized in soft fruits and leaves in the wooded areas.

The first members of our genus, *Homo*, appeared in Africa about this time, so we have the extraordinary concept of several human species living side by side. All modern humans, *Homo sapiens*, are one species – not for reasons of political correctness, but based on biological evidence. Generally, members of a species all look rather similar, but some mammalian species show considerable variation in form. The key test of species uniqueness is that members of a species can all interbreed and produce viable offspring, the so-called *biological species concept*. This is why we know that all domestic dogs, even through they may be as wildly different as a Chihuahua and a Great Dane, are members of one species. Likewise, all modern humans can interbreed and produce perfectly healthy children.

Modern humans, the genus *Homo*

The leap to modern human brain sizes only came with the origin of a new human genus, *Homo*. The first species, *Homo habilis*, lived in Africa from 2.4 to 1.5 million years ago, and had a brain capacity of 630–700 cc in a body only 1.3 metres tall. *H. habilis* may have used tools. The first fossils of *H. habilis* were found in 1960 by the famous palaeoanthropologist Louis Leakey. His wife Mary Leakey had found the human tracks in volcanic ash, as well as numerous other fossils from Africa. Their son Richard Leakey found the most complete skeleton of a similar form by the banks of Lake Rudolph (now Lake Turkana), and these have been named *H. rudolfensis*, a species closely related to *H. habilis*.

So far, human evolution had been happening only in Africa. But the next species, *Homo erectus*, escaped from Africa. The oldest examples are indeed known from Africa in rocks dated at about 1.9 million years ago, and similar dates have been suggested for *H. erectus* specimens from Georgia and from China. *H. erectus* had a brain size of 830–1,100 cc in a body up to 1.6 metres tall.

One of the richest sites for *H. erectus* is the Zhoukoudian Cave near Beijing in China, the source of over forty individuals of 'Peking Man'. They were found in cave deposits dating from 0.6 to

0.2 million years ago, associated with evidence for the use of fire, the use of a semi-permanent home base, and tribal life of some sort. *Homo erectus* sites elsewhere show that these peoples manufactured advanced tools and weapons, and that they foraged and hunted in a cooperative way. *H. erectus* in Africa perhaps made the Acheulian tools, which show significant control in their execution with continuous cutting edges all round.

Truly modern humans, *Homo sapiens*, may have arisen as much as 400,000 years ago, and certainly by 150,000 years ago, in Africa, having evolved from *H. erectus*. It seems that all modern humans arose from a single African ancestor, and that the *H. erectus* stocks in Asia and Europe died out. *H. sapiens* spread to the Middle East and Europe by 90,000 years ago.

The European story is particularly well known, and it includes a phase, from 90,000 to 30,000 years ago, when Neanderthal man occupied much of Europe from Russia to Spain and from Turkey to southern England. Neanderthals had large brains (average, 1,400 cc), heavy brow ridges, and stocky, powerful bodies. They were a race of *H. sapiens* adapted to living in the continuous icy cold of the last ice ages, and had an advanced culture that included communal hunting, the preparation and wearing of sewn animal-skin clothes, and religious beliefs. Some paleoanthropologists see the Neanderthals as distinct enough to be given their own species, *H. neanderthalensis*.

The Neanderthals disappeared as the ice withdrew to the north, and more modern humans advanced across Europe from the Middle East. This new wave of colonization coincided with the spread of *Homo sapiens* over the rest of the world, crossing Asia to Australasia before 40,000 years ago, and reaching the Americas 11,500 years ago, if not earlier, by crossing from Siberia to Alaska. These fully modern humans, with brain sizes averaging 1,360 cc, brought more refined tools than those of the Neanderthals, art in the form of cave paintings and carvings, and religion. The nomadic way of life began to give way to settlements and agriculture about 10,000 years ago.

. . . and now

The history of life has not ended. We are viewing the story from a particular timeline, and the story would have been different had this book been written by a plesiadapiform or a dinosaur. It is hard to avoid the classic narrative form in such an account. The earliest story tellers realized you must have a hero, who goes on a quest, faces untold challenges, and eventually succeeds in reaching his goal. Perhaps books about the history of life look like such a narrative, with a series of ever-more complex organisms emerging from the primeval slime, shaking off their competitors *en route*, and conquering the environment to emerge triumphant and in control of the Earth.

The record of human evolution seems to show an ever-quickening pace of change. Major innovations have occurred in succession: bipedalism (10–5 Myr), enlarged brain (3–2 Myr), stone tools (2.5 Myr), wide geographic distribution (2–1.5 Myr), fire (1.5 Myr), art (35,000 yr), agriculture and the beginning of global population increase (10,000 yr). The rate of population increase was about 0.1 per cent per annum at that time, rising to 0.3 per cent per annum in the eighteenth century, and about 2.0 per cent per annum today. In other words, the total global human population will more than double during the lifetime of any individual born today. In numerical terms at least, *Homo sapiens* has been spectacularly successful.

Evidence that the history of life is not a classic fictional narrative, however, is threefold.

First, *evolution is not teleological*. It is a fallacy to compare the evolution of life to a journey. Humans plan their journeys and have a goal in mind. Evolution cannot work that way. Evolution

165

works for the moment, selecting mercilessly which sibling survives, and which is thrown from the nest. The detailed criteria that worked in favour of sibling A last year might work against that sibling this year. A change in rainfall patterns, the death of a particular tree, a chance visit to the nest by a snake, or a new virus could change everything. Then, it may be entirely different the year after. Natural selection and fitness are relative, not absolute.

Second, *evolution has not stopped*. Evolution continues today as it always has; species arise and species become extinct. Human beings are affecting the Earth and the remainder of life in a more profound way than any species before. There is no evidence that when *Homo sapiens* has gone, everything will fall to pieces; probably quite the reverse in fact.

Third, *cockroaches are the pinnacle of evolution* – to other cockroaches. We might like to regard ourselves as the most successful species on Earth because we occupy so much of the Earth's surface, and control so many million square kilometres of farmland. But there are probably more cockroaches than humans. And, taken further, there are certainly more bacteria and other microbes than humans. We can define ourselves as the most successful species on Earth by careful choice of the terms by which that decision is made. Doubtless a sapient cockroach would write a different book.

Index

A

age of the Earth 16–20
Altman, Sidney 26
amphibians 79, 85, 97, 109–11,
    129–30
archosaurs (ruling reptiles)
    130–1, 134, 137
anthropods 77–9, 82–3, 85, 156
apes and primates 9–10, 147–9,
    152–4, 157–9, 160–2
Archaea 35, 37, 39, 41
arthropods 60–1, 65
australopithecines 161–3

B

bacteria 30, 35, 37, 39, 41
Bambach, Dick 102
bangiomorpha 45–7
Becquerel, Henri 17
bipedalism 159–62
biochemical theory 23–5
brain volume 159, 162–3
Brocks, Jochen 37
Brunet, Michel 161
Buick, Roger 23
Burgess Shale, Canada 5, 8, 61,
    63, 65
Butterfield, Nick 44–5

C

Cambrian explosion 50, 51–2,
    56–68
Canada 5, 8, 30, 61, 63, 65, 98–9
carbon 20–1, 22, 40, 85–100
Carboniferous period 85–100

Carnian crisis 132–3
Carson, Rachel 69
Catling, David 40
Cech, Thomas 26
Chengjiang fauna, China 59–63
China 5, 8, 59–63, 70–1, 106–7,
    112–15, 138, 163–4
Chordata 59, 60, 61–3
Clack, Jenny 81, 82
cladistics 11–13
climate 40–1, 89, 116–19, 123–4,
    154
coal 87, 91–5
Coates, Mike 81, 82
continents 88–9, 108–9
coral 51–2, 59, 103–5, 125
Cretaceous terrestrial explosion
    122–3, 142–5, 147, 151,
    155
Crick, Francis 9, 26
crocodilians 135, 142, 147
cyanobacteria 29, 38, 40–1, 71
cynodonts 130–1, 136

D

Dart, Raymond 162
Darwin, Charles 6–7, 8, 16, 51, 52
Dawson, William 98
dinosaurs 1–6, 122, 126–40, 142,
    144–8
    Carnian crisis 132–5
    extinction 144–5, 147
    KT event 122–3, 144–5, 147, 151, 155
    Mesozoic era 132–40, 144–5, 147,
    151, 155
DNA (deoxyribonucleic acid)
    8–10, 25–6, 44, 66,
    158–9
Doushantuo, China 70–1

E

Earth 16–21
echinoderms 57–8, 63

Ediacaran fossils 49–52, 66
Edwards, Dianne 73–4
eggs 99–100, 137–8, 148
enosymbiotic theory 37–9
Eucarya 35–6, 37, 39, 41
eukaryotes 32, 34–5, 37–9, 41–2,
    44–5
extinction *see* mass extinction

F

fishes 62, 67, 79–84, 96, 126–9,
    142
five digits, significance of 81–2
fossils 1–8, 27, 29–30, 41–2, 47–8,
    61, 63, 65, 138–40
    blimps and ghost range 7–8
    Ediacaran fossils 49–52, 66
    exceptional preservation 4–5, 8
    first fossils 28–32
    Grypania fossils 41–2
    humans, origins of 161–4
    microfossils 28–32
    plants 74–6
    reptiles 98–100
    Rhynie fossils 74–6, 78
    stromatolites 28–30
Fox, Sidney 25
fungi 35, 36, 70–1

G

genes 25–7
Gilbert, Walter 26
Gondwana 108
Gould, Stephen 65
Gunflint Chert of Canada 30

H

Haldane, JBS 23–5
Hallam, Tony 106–7
Hennig, Willi 11
Hoffman, Paul 48
Holmes, Arthur 17, 20

homo 164–6
humans, origin of 9–10, 81–2,
    146–66
Huxley, Thomas Henry 41
Hylonomus 99–100

I

ice caps 123–4
insects 65, 70, 77–8, 87, 93, 143–4
isotopes 92, 115–20

J

Joggins Cliff, Canada 98–9
Johanson, Don 162
Johnston, Eric 122

K

Karoo Basin, South Africa 109
Kirschvink, Joseph 48

L

land, origins of life on 69–86,
    109–12
Laurasia 108
Leakey, Louis, Mary and Richard
    163
lemurs 153
Liaoning Province, China 5, 8,
    138
lichens 70–1
lizards 10, 135, 142
'Lucy' 162
Lyell, Charles 98
Lystrosaurus 111–12, 123–4

M

MacNaughton, Robert 77
mammals 10, 79, 100, 130, 136,
    148–55

Margulis, Lynn 37
marsupials 149, 155–6
mass extinction 101–21, 144–5
   background extinction 101
   'big five' mass extinctions 101–2
   climate change 116–19
   continental drift 108–9
   dinosaurs 144–5, 147
   disaster species 113
   end-Permian event 102–15
   KT event 144–5, 147, 151, 155
   land, life on 109–12
   oxygen 106–8, 116, 117, 119
   recovery 119–21, 122, 129–31, 139
   reptiles 111–12, 145
   rivers and shock erosion 115–16
   runaway greenhouse phenomenon
     118–19
   sea, life in the 103–6, 125–6
   Siberian Traps, vast volcanic
     eruptions in 108–9, 118–20
   soil erosion and sediment run off
     115–16, 118–19
   timing of event 112–15
   Vyatskian assemblage, Russia
     109–11
Maynard Smith, John 33–4
Meishan, China 106–7, 112–15
Mesozoic era 122–45, 151, 154–5
Miller, Stanley 24, 26
mitochondria 35, 38–9
modern ecosystems, origins of
   122–45
Mojzsls, Stephen 22
molecular biology 8–14, 66–8
molecular clock 8–10, 66–8
molecules 149–52
monkey-rabbits 152–3
monkeys 156–8
morphology 149–52
multicellularity 42–3, 44–5, 48

N

Neanderthals 164
Nwell, Andy 16

Nisbet, Euan 25, 30

O

Oparin-Haldane model 23–6
origin of life 15–32
oxygen 6, 20–1, 39–41, 47–8, 64,
   91–3, 96, 100, 106–8,
   116, 118–19

P

Pangaea 108, 123–4
Pasteur, Louis 15–16
Pauling, Linus 8–9
Peterson, Kevin 67
phylogenetic argument 44–5
placental mammals 149, 150–2
plants 5, 34, 59–63, 71–7, 94–5,
   111–12, 118–20, 124,
   135–6, 143–4
predation 141–2, 147
primates 147–9, 152–4
prokaryotes 35, 37–9, 41–2,
   44

R

radioactivity and radiation 17,
   40
Raup, David 101
reptiles 79, 85, 97–100, 109–12,
   122–3, 130–1, 134,
   136–7, 145
Rice, Clive 75
rivers and shock erosion 115–16
RNA (ribonucleic acid) 25–8,
   66
rock dating 14, 17, 20, 21–3, 52
rodents 152–3
Romer, Alfred Sherwood 83
runaway greenhouse
   phenomenon 117–19
Rutherford, Ernest 17–18, 20

Index

S

Sarich, Vincent 9
sclerites 56-9
sea, life in the 87-8, 103-5, 112-29, 140-2 *see also* fishes
Seilacher, Dolf 50, 51-2
Selden, Paul 77
Senut, Brigitte 161
sex, origin of 32, 33-50, 85
sharks 87-8, 142
Siberian Traps, vast volcanic eruptions in 108-9, 118-19
skeleton, origins of the 51-68
Sleep, Norman 25, 30
Small Shelly Fauna (SSF) 54-6, 57
Smith, Roger 115
Snowball Earth (Cryogenian) 47-8
Soddy, Frederick 17
South Urals, Russia 109-10
Sprigg, Reginald 49
Springer, Mark 149-50
stromatolites 28-30
Sun, formation of 20-1

T

tetrapods 79-85, 96-100, 123, 142
theropods 134, 138
Thomson, William (Lord Kelvin) 16-17

tree of life 11-14, 34-7, 39, 41, 45
Trewin, Nigel 75
trilobites 58-61, 65
Tverdokhlebov, Valentin 115
Twitchett, Richard 107-8, 117, 125

U

Urey, Stanley 24

V

Van Valen, Leigh 121
Vyatskian assemblage, Russia 109-11

W

Ward, Peter 115-16
Watson, James 9
Wellman, Charlie 72
Wible, John 151-2
Wignall, Paul 106-7, 117
Wills, Matthew 65
Wilson, Allan 9
Woese, Carl 35
worms 2, 4-6, 7, 50, 66, 79
Wray, Greg 66-7

Z

Zhoukoudian Cave. China 163-4
Zuckerkandl, Emil 8-9

Expand your collection of
VERY SHORT INTRODUCTIONS

1. Classics
2. Music
3. Buddhism
4. Literary Theory
5. Hinduism
6. Psychology
7. Islam
8. Politics
9. Theology
10. Archaeology
11. Judaism
12. Sociology
13. The Koran
14. The Bible
15. Social and Cultural Anthropology
16. History
17. Roman Britain
18. The Anglo-Saxon Age
19. Medieval Britain
20. The Tudors
21. Stuart Britain
22. Eighteenth-Century Britain
23. Nineteenth-Century Britain
24. Twentieth-Century Britain
25. Heidegger
26. Ancient Philosophy
27. Socrates
28. Marx
29. Logic
30. Descartes
31. Machiavelli
32. Aristotle
33. Hume
34. Nietzsche
35. Darwin
36. The European Union
37. Gandhi
38. Augustine
39. Intelligence
40. Jung
41. Buddha
42. Paul
43. Continental Philosophy
44. Galileo
45. Freud
46. Wittgenstein
47. Indian Philosophy
48. Rousseau
49. Hegel
50. Kant
51. Cosmology
52. Drugs
53. Russian Literature
54. The French Revolution
55. Philosophy
56. Barthes
57. Animal Rights
58. Kierkegaard
59. Russell
60. Shakespeare
61. Clausewitz
62. Schopenhauer
63. The Russian Revolution
64. Hobbes
65. World Music
66. Mathematics
67. Philosophy of Science
68. Cryptography
69. Quantum Theory
70. Spinoza
71. Choice Theory
72. Architecture
73. Poststructuralism
74. Postmodernism
75. Democracy
76. Empire
77. Fascism
78. Terrorism
79. Plato
80. Ethics
81. Emotion
82. Northern Ireland
83. Art Theory
84. Locke
85. Modern Ireland
86. Globalization
87. Cold War
88. The History of Astronomy
89. Schizophrenia
90. The Earth
91. Engels
92. British Politics
93. Linguistics
94. The Celts

95. Ideology
96. Prehistory
97. Political Philosophy
98. Postcolonialism
99. Atheism
100. Evolution
101. Molecules
102. Art History
103. Presocratic Philosophy
104. The Elements
105. Dada and Surrealism
106. Egyptian Myth
107. Christian Art
108. Capitalism
109. Particle Physics
110. Free Will
111. Myth
112. Ancient Egypt
113. Hieroglyphs
114. Medical Ethics
115. Kafka
116. Anarchism
117. Ancient Warfare
118. Global Warming
119. Christianity
120. Modern Art
121. Consciousness
122. Foucault
123. Spanish Civil War
124. The Marquis de Sade
125. Habermas
126. Socialism
127. Dreaming
128. Dinosaurs
129. Renaissance Art
130. Buddhist Ethics
131. Tragedy
132. Sikhism
133. The History of Time
134. Nationalism
135. The World Trade Organization
136. Design
137. The Vikings
138. Fossils
139. Journalism
140. The Crusades
141. Feminism
142. Human Evolution
143. The Dead Sea Scrolls
144. The Brain
145. Global Catastrophes
146. Contemporary Art
147. Philosophy of Law
148. The Renaissance
149. Anglicanism
150. The Roman Empire
151. Photography
152. Psychiatry
153. Existentialism
154. The First World War
155. Fundamentalism
156. Economics
157. International Migration
158. Newton
159. Chaos
160. African History
161. Racism
162. Kabbalah
163. Human Rights
164. International Relations
165. The American Presidency
166. The Great Depression and The New Deal
167. Classical Mythology
168. The New Testament as Literature
169. American Political Parties and Elections
170. Bestsellers
171. Geopolitics
172. Antisemitism
173. Game Theory
174. HIV/AIDS
175. Documentary Film
176. Modern China
177. The Quakers
178. German Literature
179. Nuclear Weapons
180. Law
181. The Old Testament
182. Galaxies
183. Mormonism
184. Religion in America
185. Geography
186. The Meaning of Life
187. Sexuality
188. Nelson Mandela
189. Science and Religion
190. Relativity
191. History of Medicine
192. Citizenship
193. The History of Life

DARWIN
A Very Short Introduction
Jonathan Howard

Darwin's theory of evolution, which implied that our ancestors were apes, caused a furore in the scientific world and beyond when *The Origin of Species* was published in 1859. Arguments still rage about the implications of his evolutionary theory, and scepticism about the value of Darwin's contribution to knowledge is widespread. In this analysis of Darwin's major insights and arguments, Jonathan Howard reasserts the importance of Darwin's work for the development of modern science and culture.

> 'Jonathan Howard has produced an intellectual *tour de force*, a classic in the genre of popular scientific exposition which will still be read in fifty years' time'
>
> **Times Literary Supplement**

www.oup.co.uk/isbn/0-19-285454-2

PSYCHOLOGY
A Very Short Introduction
Gillian Butler and Freda McManus

Psychology: A Very Short Introduction provides an up-to-date overview of the main areas of psychology, translating complex psychological matters, such as perception, into readable topics so as to make psychology accessible for newcomers to the subject. The authors use everyday examples as well as research findings to foster curiosity about how and why the mind works in the way it does, and why we behave in the ways we do. This book explains why knowing about psychology is important and relevant to the modern world.

'a very readable, stimulating, and well-written introduction to psychology which combines factual information with a welcome honesty about the current limits of knowledge. It brings alive the fascination and appeal of psychology, its significance and implications, and its inherent challenges.'

Anthony Clare

'This excellent text provides a succinct account of how modern psychologists approach the study of the mind and human behaviour. ... the best available introduction to the subject.'

Anthony Storr

www.oup.co.uk/vsi/psychology